LONGMAN
INTERNATIONAL
TECHNICAL
TEXTS

BLOCKLAYING AND CONCRETING

Second Edition

M.O.Obande

Head of Department, Department of Building and Dean of Environmental Design Kaduna Polytechnic

LONGMAN

This book is dedicated to my wife,
Enotse, who typed the manuscript.

Pearson Education Limited
Edinburgh Gate, Harlow,
Essex CM20 2JE, England
and Associated Companies throughout the world

First Published 1981
Second Edition 1990
Sixth impression 1999

Printed in Singapore (COS)

ISBN 0-582-02541-9

Preface

Long experience in the teaching of building subjects, both at the craft and technician levels has led to the writing of this book. While there are many specialised books on brickwork, there are very few specialised books on blockwork. I have therefore felt that this small book on blocklaying and concreting may not only help the beginner in understanding the basic needs of his new craft but also fill the gaps left between brickwork and blockwork.

Blockwork has become so popular in many parts of the developing world that, although similar to brickwork in many respects, it deserves a separate and greater emphasis. The West African Examinations Council, for example, changed from examining students in Brickwork to Blocklaying and Concreting to reflect the extensive use of blockwork in West Africa.

Although I have written mainly for craft students, particularly those taking full-time and part-time courses in technical schools, colleges and craft schools, I hope that students taking building and architectural courses in polytechnics and colleges of technology will also find the book valuable. Old craftsmen too will find the book an interesting source of reference.

The book covers the City and Guilds of London Institute's Intermediate Syllabus in Blocklaying and Concreting and other local syllabuses.

Every topic discussed has been clearly illustrated with diagrams to aid the beginner in understanding the necessary craft techniques involved. Each chapter ends with revision questions to help the reader assess his understanding of the chapter.

I would like to acknowledge the help and encouragement I have received from Mr A.O. Echebiri of Longman Nigeria Limited, Mr Rob Francis and Dr Katie Petty-Saphon of Longman Group UK Limited and also the friendly advice and criticism from my many friends in the building profession, particularly M.Y. Javaid, R.O.A. Osha and J.A. Gawati all formerly of the Kaduna Polytechnic, Department of Building Technology.

In this second edition two new chapters have been added. These chapters have been designed specifically to teach the practical aspects of the trade in a progressive manner so as to help correct the student failure rate in practical examinations. This section of the book simplifies the job of the instructor by presenting him with a good number of prepared practical models for the weekly workshop practices for his class. The student is gradually led towards the ultimate objective of his course through the graded exercises available to him in the book for constant shop practice.

I should like to acknowledge the numerous requests from instructors and students to include a section on practicals in this edition. Considerations of space have limited the number of such exercises included, nevertheless the exercises will no doubt be of great help to the reader.

M.O.O.

Acknowledgements

The Publishers are grateful to the following for their permission to reproduce photographs:

Cement and Concrete Association for Fig. 180; J. Allan Cash for Fig. 1; Katie Petty-Saphon for Figs. 2, 25, 51, 86, 93, 100, 101, 133, 145, 155, 159, 179, 186, 187, 188, 191 and 197 and Tropix Photo Library for Fig. 3 (Martin Birley) and Fig. 135 (D. Drakakis-Smith).

All other photographs not acknowledged were supplied by the author.

We are also grateful to the following for permission to reproduce past examination questions:

City and Guilds of London Institute for question from *Brickwork Craft – Practical* May–June 1974, paper no. 608-1-03; West African Examinations Council for questions from *Blocklaying and Concreting, Intermediate Craft* Dec 1973, May 1978, 1981, 1984, paper nos. 831-0-03, 898-1-03 and *Blocklaying and Concreting, Advanced Craft* May 1979, 1980, 1984, paper nos. 831-0-12, 898-2-12.

Contents

1. Traditional walling materials

Most forms of building are traditionally based on locally produced materials. This is partly due to the high cost of transporting heavy building materials for long distances, and partly due to the climatic differences that exist in different parts of the world.

What is considered traditional varies from one place to another and from period to period. In many communities in the past, buildings were erected through joint efforts by the local community, using locally and easily obtainable materials. The provision of dwelling houses was regarded as the responsibility of the community. There was no monetary payment for such labour, but the owner provided food for the workers as long as they remained on the job.

Most traditional building materials are poor in durability and therefore have to be more frequently replaced than modern building materials.

Traditional architecture as far as we are concerned, can be classified into two groups:

1 domestic architecture, which survives mainly in the rural areas now,

2 monumental architecture, used for the construction of religious buildings and shrines.

It is regrettable that many traditional architects have so far ignored researching into their architectural past.

Fig. 1 An example of a traditional house in Nepal. The wall is formed of earth blocks and the roof is of neatly made thatch

Traditional building materials

The more commonly used traditional materials include stone, clay and mud, and timber.

Stone

Stone is not traditionally a popular material for the construction of walls, for obvious reasons. It is difficult to obtain in certain areas, and in those areas where it is obtainable, it is difficult to shape into sizes that can be handled with ease. There is no traditional jointing material except mud. The construction of dwelling houses with stones is therefore hazardous.

Stone is, however, a valuable material for the formation of bases of walls.

Fig. 2 A typical example of the mud buildings for which the ancient city of Kano, Nigeria is famous. Note that the horned, domed roofs are also formed of mud and wood

Perhaps its commonest use can be found in the formation of foundations for rural buildings. In areas where stone is plentiful, forms of construction exist which are based on relatively thick walls made of loosely piled stones gathered from the vicinity of the building.

Clay and mud

These materials are extensively used in the construction of traditional buildings. The use of clay and mud takes various forms. The clay or mud can be excavated, kneaded, and built into shape to form the required thickness for a wall. A mixture of clay and straw obtained by treading, and known as 'cob', is also used for the construction of earth walls. This mixture is usually moulded into small units or bricks, dried and later built to form a wall.

In areas where timber is plentiful, a mixture of semi-dry clay, chalk and gravel is used for walls by ramming and consolidating it between a timber framework. Similarly, in areas where a special type of grass with hard, smooth, cane-like stems is available, the grass is woven into mats, tied to reinforcing bamboo or similar poles and finally covered with a daub of cheaply available local material, such as clay, mud or cattle waste.

Some of these forms of construction are more labour-consuming than others and some need more skilled labour than others. As most houses are self-built or built by the community, the construction takes place at a time when the labour would otherwise be unoccupied.

Structurally, load-bearing walls of earth are quite satisfactory for a small structure providing that the walls are kept dry. They are adequate for ordinary dwelling purposes and provide good insulation and a comfortable internal temperature.

Timber

In areas where timber is plentiful, timber framed and often boarded constructions are common, logs of timber being arranged to form the walls of dwelling houses. Bamboo may be similarly used in areas where it is available.

Extensive use of timber is what would be expected in areas with abundant forest products,as timber buildings satisfy most of the economic and physical conditions required for ordinary housing needs. Timber houses may, however, be regarded as belonging to the poor who cannot afford to build with cement blocks. This is despite the advantages associated with timber construction some of which are as follows.

1 It is cheap.
2 It is easily handled by non-regular workers.
3 It provides a flexible form of construction which can be easily repaired, and is adequate for protection against sun and rain in countries with hot humid climates.

The major disadvantage of the use of timber for construction of walls in tropical countries can be the high cost of eradicating termites. Termites will destroy structural timber unless adequate precautions are taken. In addition, timber is highly flammable and hence unsuitable for urban areas where buildings are very close together.

The durability of traditional earth buildings

Shrinkage, erosion and underscouring are particularly common problems with earth walls. These problems are made worse if the building:

1 is in a position where it is exposed to direct sunlight,
2 is made of clay,
3 is subjected to well-defined periods of sunshine and rainfall.

The following precautions may be taken to prevent excessive deterioration:

1 addition of grass, coconut fibres, coarse sand, etc. to the earth during mixing,
2 use of wide eaves (the overhanging edge of the roof), to prevent direct rainfall on the walls,
3 choosing a site on high ground, in a sheltered area to retain moisture within the walls,
4 rendering the external surfaces of the walls with water-repellent materials, such as a mixture of bitumen, cement, etc.

Essential functions of a building

Every building should fulfil the following functions:

1 protect human inhabitants, animals, materials and equipment from the effects of weather, e.g. rain, sunshine and wind,
2 provide adequate internal comfort for the inhabitants, e.g. a healthy temperature and dry environment,
3 be durable.

The last function can be achieved if adequate steps are taken right from the setting out and the construction of the building. Foundations should be deep enough for the type of building envisaged, and be formed of inert and durable materials. Workmanship is as important to the durability of a building as the quality of the materials. Often the weakness of block walls lies with the joints. Cement mortar should be mixed in small quantities at a time to prevent it becoming dried out by the sun before it has been used to make joints. The quantity mixed will depend on the number of users. Walls built the previous day and the blocks to be used during the day should be soaked with water before the new work is begun.

For roofs and roofing materials, the precautions necessary to ensure durability depend upon the type of material being used. If the roof is of wood, the quality of the wood, e.g. adequate seasoning, the correct type of wood, etc. must be ensured before use. As a roof is formed of various members, the joints are vital in order to achieve reasonable rigidity. Concrete roofs should have enough reinforcing metal, which must be placed at the correct positions ascertained by the designer. Similarly, the materials for making the concrete–stone, sand and cement–should be of the best quality. Reference should be made for example to your country's Standard Code of Practice with regard to the quality of materials for structural concrete.

Every building should have walls thick enough to prevent collapse on impact, vibration from vehicles and wind force. Walls act as barriers to the spread of fire. Dwelling houses in developing countries tend to be overcrowded with consequent greater fire risk. This fact makes it all the more important to consider wall thicknesses not only in the light of providing enough support for the roof but also for general safety purposes.

Revision questions

1 Name the three types of traditional building materials most common in your area.
2 List the advantages and disadvantages of using soil as a walling material.
3 Why is stone not a popular traditional walling material?
4 List four precautions to be taken to limit the rapid deterioration of earth walls.
5 What are the main functions of a building?
6 What steps should be taken when using the following materials to ensure that they perform their correct functions in the building:
a) cement mortar,
b) wood,
c) hollow concrete blocks?
7 Specify the necessary precautions to be taken when using sand/cement mortar during dry weather to ensure its durability.

2. Parties concerned with the construction of a building

A good craftsman should not only know the practical aspects of his trade but also the administrative aspects. The administration of building contracts can be grouped under two broad headings: direct labour administration, and contracted administration. The former is carried out by the owner himself, while the latter is done by a building firm.

The use of terms to describe the different personnel concerned with a building contract seems to be in a confused state in many parts of the world. This confusion may be as a result of the diversity of building firms operating in the industry. A general foreman in one firm may be a contracts manager in another and a site agent in yet another. The qualifications and necessary experience of supervisory personnel within the industry have yet to be clearly defined. We shall therefore here attempt to give definitions of these parties as they are most commonly used.

Contractor's party

The contractor or builder

This is the person or group of persons who undertakes to use his employees, plant and materials to construct a building according to drawings and specifications provided by the client or building owner. Building contracts are normally carried out under a legally binding contract entered into by the builder and the building owner.

Contracts manager

This post may only exist in medium to large building firms. His functions are mainly administrative. He takes charge of all the contracts within his area of operations. He sees that progress is kept at a predetermined pace throughout the duration of the contract.

General foreman

The general foreman is the person directly in charge of site work. He is responsible to the contracts manager for orderly progress of the site work. He has under his control the various trades foremen. He usually attains his position by a series of promotions through the ranks of a craftsman, either as a blocklayer or a carpenter. Nowadays, due to the complexity of building operations and with the advent of numerous services and design features, the general foreman is expected to have attained relevant technical qualifications, such as a Full Technological Certificate in Building Crafts of the City and Guilds of London Institute or an equivalent certificate. A general foreman should be able to criticise intelligently the standard of workmanship of all trade operations on the site.

Trades foremen

They are responsible for explaining the daily work allocations to their workers. They supervise the speed and workmanship of their different craftsmen and alert the general foreman to any deviation from the planned programme, in order that corrective measures may be taken early enough.

Gangleader

On a large job where the main contractor carries out all the building work, such as blocklaying, concreting, carpentry, plumbing, painting, etc., the number of craftsmen in each trade becomes too large for effective control by one foreman. The craftsmen are therefore divided into gangs, each gang being controlled by a gangleader. The gangleader or ganger, as he is sometimes called, carries out his normal operations as a craftsman, and in addition he maintains order and harmony among the workers in his gang. He subdivides the gang's allocation of the job between the individual members and keeps note of all overtime done by the gang or by individual members.

Timekeeper

The timekeeper is an important person in a building contract. Most building contracts are carried out on open sites and, for economic reasons, most sites are not fenced in. This gives dishonest workers an opportunity to come to work late using accesses other than the main entrance, thereby cheating the contractor of a full day's pay. To some extent, a profit or loss on a contract may depend on the attitude of the timekeeper. In some sites the timekeeper is also the storekeeper. An honest timekeeper is therefore a valuable asset to the contractor.

Fig. 3 A gateman at work guarding the entrance to a construction site

Gateman

The gateman is useful in preventing unauthorised entry into the site by men and vehicles. In some areas, building sites provide an opportunity for local do-it-yourself builders to steal the contractor's materials for their own work. Tools and equipment, especially buckets, headpans, shovels and pick axes are well known targets for thieves. It is also common knowledge that some storekeepers and foremen cheat the contractor by collaborating with subcontractors who deliver materials to the site and leave again without discharging their loads. Although most gatemen are old and lack the strength needed to fulfil their functions satisfactorily, the mere sight of them at the gate discourages intending criminals.

Client or building owner's party

The client

Through his architect and quantity surveyor the client provides all the necessary information in the form of drawings, specifications, charts, schedules, etc. for the construction of the building. He also provides the funds for the payment of the contractor on completion of the contract or whenever payment is required under the terms of the contract. He also pays for services employed to serve his interest in the contract. The client may be an individual, group of persons, government or other organisation.

Supervising architect

He is employed by the client to see that the contractor complies with the drawings and specifications and the Codes of Practice in executing the work. The architect accomplishes this task by delegating the duty to a clerk of works whose office must be on the site.

Clerk of works

The use of the clerk of works to supervise building contracts is becoming increasingly popular in the industry. This may be due to the availability of suitable people graduating from colleges of technology and polytechnics. As pointed out above, the responsibilities of the clerk of works involve the checking of materials and levels, setting out and the testing of materials to ensure compliance with standards approved by the architect.

Quantity surveyor

The client may also appoint a quantity surveyor whose duties will be:

1 measuring and determining the value of the work done,
2 measuring and keeping a check on authorised variations that may occur in executing the contract,
3 checking and authorising interim and final certificates for payment.

Statutory personnel

These are members of the supervisory staff whose offices are established by statutory instrument. They include the following personnel.

Building Inspector

He is an employee of the local authority who sees that local building laws are enforced on the site.

Public Health Inspector

He sees that health regulations are kept on the site by the contractor.

Inspector of Factories

He checks on the safety of men and equipment on the site.

Revision questions

1 Name the various supervisory personnel involved in the construction of a building.
2 Define the following, giving an example in each case:
   a) contract work,
   b) direct labour.
3 To whom is the clerk of works responsible?
4 What are the main duties of the following site staff:
   a) timekeeper,
   b) trade foreman,
   c) gangleader,
   d) clerk of works?
5 What are the duties of the Inspector of Factories on a building site?

3. Preliminary site operations

Site investigation and examination

In major building operations, the investigation of the site involving the soil, type of terrain, topography, obstructions, subsoil installations, proximity to existing properties, etc. is carried out at the design stage by the designing group. For the purpose of tendering for such a contract, the contractor, in his own interest, also carries out these aspects of site investigation.

Most buildings are not large enough to necessitate elaborate site investigation prior to commencement of operations. It is, however, essential and advisable on the part of the builder to examine the proposed site before building operations start in order to ensure that some statutory regulations are not contravened.

Every contractor should take appropriate steps on the award of the contract to satisfy himself of the following.

1 There are no local authority restrictions on methods of work within the area, e.g. level of noise.

2 The site does not slope steeply. If it does, the volume of cut and fill necessary must be determined.

3 If there are existing structures to be removed, the method of removal envisaged is appropriate and constitutes no hazard to anyone.

4 The building is set out correctly, observing the building line for the area.

5 A tracing of subsoil services existing in the area is obtained from the land survey department of the local authority.

Advantages of site investigation

1 It saves the contractor from future legal tussles with the local authority and/or occupiers of adjacent property.

2 It forewarns the contractor of the type of problem to be expected during the operations and enables him to take steps in advance to solve them.

3 It prevents unnecessary danger to workmen, through proper planning of the operations.

Site clearance

Site clearance is necessary for all types of buildings. If the new building is to be built on an area where a building already exists, the old building has to be demolished, the debris cleared away and existing services disconnected and removed. Demolition does not come within our present level. Virgin sites have to be cleared of grasses, trees, rocks and old vehicles. The disposal of this rubbish may constitute a problem to the contractor in heavily populated and congested city sites. Timber from felled trees, grasses and shrubs can be burnt. Permission should be obtained from the local authority if much has to be burnt. Agreement should be reached as to the best time of the day to do the burning, and notification should be sent to the office of the town's fire brigade. Roots of trees should be properly removed to ensure that they do not cause structural damage to floors, pavements and roads when rotten.

In towns, some trees may have preservation orders on them. It is an offence to destroy such trees and the contractor can be heavily fined for the offence. Enquiries should be made to the town planning office to ascertain that trees with preservation orders do not exist within the proposed site before clearance is started.

Site clearance used to be carried out using manual labour with pick axes and spades. Due to the mechanisation of building operations, there are now mechanical plants which can carry out site clearance operations with comparative ease and safety. These plants are expensive to purchase or hire and only medium and large sized firms can afford to own them. Site clearance is therefore still largely done using manual labour, especially for housing projects.

Site levelling

After clearance, the site has to be levelled. Many building sites were originally parcels of land used for cultivation of farm crops, and often contain mounds and ridges. These have to be levelled to assume a common surface before the setting out of the building can be accomplished. Nowadays, mechanical plants are invariably used for the purpose, except where the building is a small domestic type. This stage of levelling is usually referred to as 'reduced level' and it involves cutting high areas of the soil and filling up the lower areas within the site.

Setting out and fixing levels

In building contracts, the planning of the project and its design are often done before the builder or contractor is selected. This means that the contractor will have the task of interpreting the design provided by the designer. This is why setting out of the building demands careful consideration right from the earliest stages of the contract. Setting out deserves the attention of an expert, as mistakes or a badly done job at this stage may prove too costly for the contractor to amend at a later stage.

Correct setting out should ensure that:

1 the building faces in the right direction,

2 the building is erected on the correct plot of land,

3 the overall sizes (length and breadth) are according to those shown on the drawing,

4 regard is paid to the local building line.

Although setting out is a one-off operation for a simple building, in a more complicated building, it is a continuous operation throughout the construction of the building and it can dictate the progress of a project. At this level, we shall concern ourselves mainly with the basic principles of setting out simple buildings.

Building contracts almost always start in a rush. The owner of the building often wants to see his building completed at the earliest possible date after the award of the contract. The contractor, who might not be well acquainted with details of the design, starts the work in a rush. It therefore often happens that important items of work, such as site investigation, checking and recording site boundaries, party walls and bench marks, are neglected. Such oversight usually results in costly remedial action by the contractor later in the contract.

There are several reasons for this trend, for example shortage of staff, lack of information, reluctance on the part of the contractor to employ the right type of staff, or failure to appreciate the value of correct setting out.

The number of public services is increasing rapidly in modern buildings and this further complicates the job of setting out. Setting up the site must be considered at the time of setting out. It is not sufficient just to obtain information from the statutory suppliers of water, electricity, telephone, etc. The information they give will only be approximate, so site layout is an essential task in the interest of safety and of future building progress.

Building line

A building line, which is fixed by the authority, is a statutory line in terms of a dimension beyond which no building may project. It normally fixes the front line of all buildings along a road or street. Before any setting out is commenced, the position of the building on the site must be determined. This is found from the site plan which refers only to the position of the building and gives no detailed information as to the building itself. The position of the building line is usually indicated by a dimension from a fixed line or point such as an existing building or roadway.

Datum peg

Before starting the actual setting out of the building, it is essential to establish a level on the site, to which references pertaining to levels of elements of work may be made in the course of executing the work. In planned cities and towns Ordnance Bench Marks are established and the points marked along streets in permanent positions, or cut into plinths of walls and gate posts or other reasonably permanent fixtures. Whenever possible, Ordnance Bench Marks should be used in connection with establishing the level of land and surface water drains, floors, etc. Where a Bench Mark is far from the proposed site, a levelling instrument such as a cowley level can be used to transfer the datum level to the site, from where it can be distributed around the proposed building.

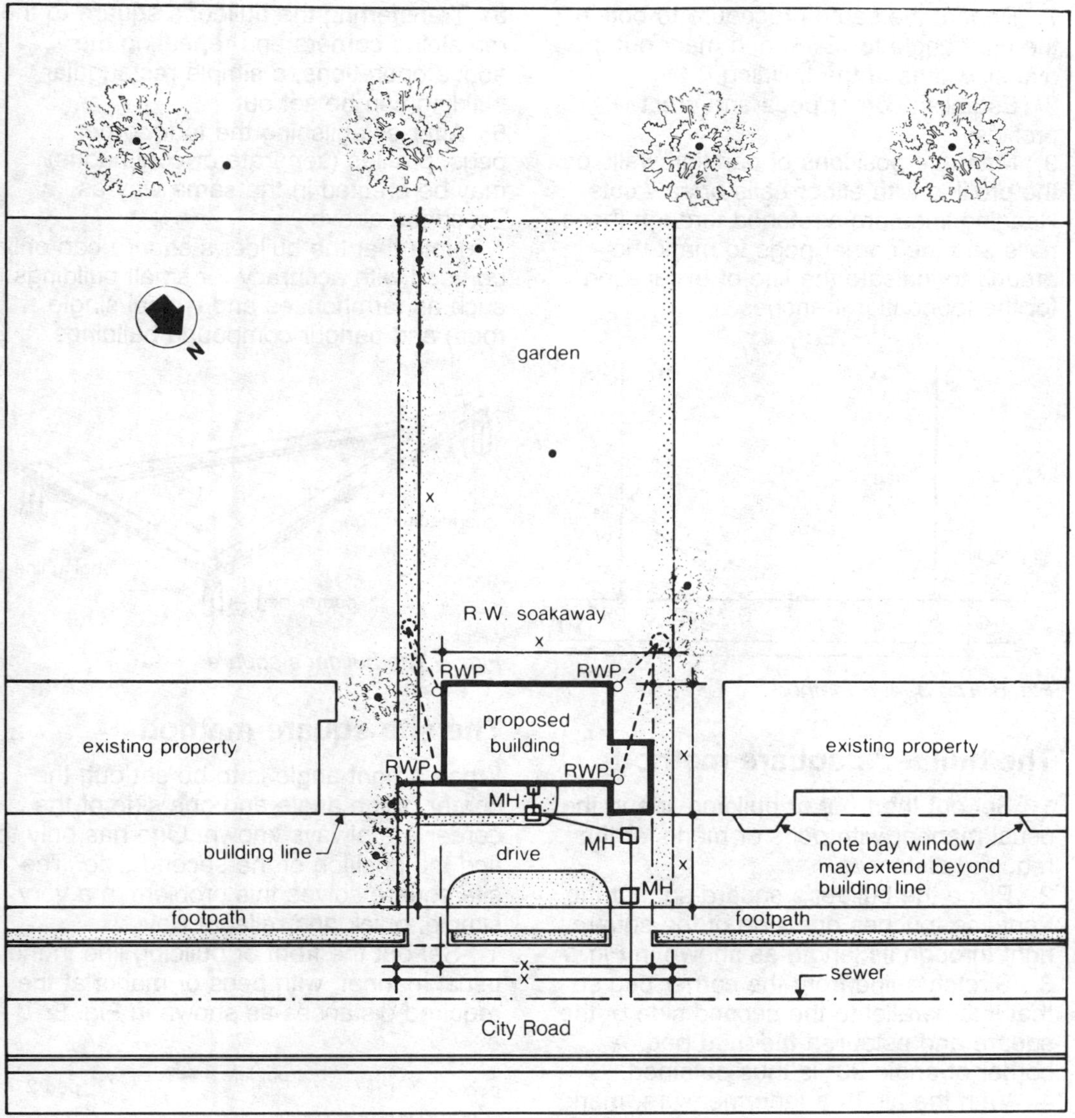

Fig. 4 A site plan

On a site where the adjacent plots of land have been developed already, the level of the existing floor, pavement, threshold or the crown of the estate road may be similarly transferred using a levelling instrument. If adequate attention is paid to the level of the structural parts of the proposed building, especially the ground floor level, in relation to the level of the surrounding land, expensive remedial work in the future will be avoided. During wet weather flood water can prove to be a menace in buildings with low-lying ground floors.

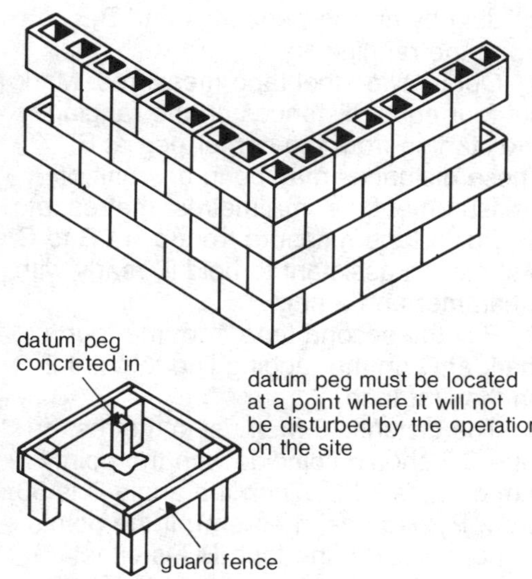

Fig. 5 The datum peg in relation to the external wall of a building

Methods of setting out

In setting out simple buildings we may use one of three methods. These are:
1 the 3, 4, 5, method,
2 the builder's square method,
3 levelling instruments.

The 3, 4, 5 method

1 Mark out the building line from the road by measuring the required distance from the centre of the road, or by stretching a line along an existing building to the proposed site. The building line is then represented by a line known as the ranging line, which also marks the front wall of the buildings, as shown by GG_1 in Fig. 6.
2 Mark out the overall length of the building by driving pegs at A and B, along the ranging line.
3 Obtain two steel tape measures. Mark out four equal distances on the ranging line starting from the corner peg at B. These distances may be in any unit of measurement, i.e. millimetres, metres, etc.
4 Pull a tape measure from point B to C and ask an assistant to hold it, ready with a hammer and a peg.
5 Pull the second tape from the fourth mark at D on the ranging line to point E on the first tape.
6 The distance 5 m, if using metres, on tape DE should coincide with the point 3 m on tape BEC. Then the angle B is 90° (from Pythagoras' theorem). If the points do not coincide, the tape BC is either shifted outwards or inwards until 5 m on the second tape coincides with the 3 m mark on the first tape.
7 Repeat the same procedure to obtain the right angle for BAF, and mark out the overall widths of the building.
8 Establish corner pegs and erect profiles.
9 Mark the positions of partition walls on the profiles with either nails or saw cuts. Ranging lines are stretched through these nails and the corner pegs to mark the ground to indicate the line of excavation for the foundation trenches.

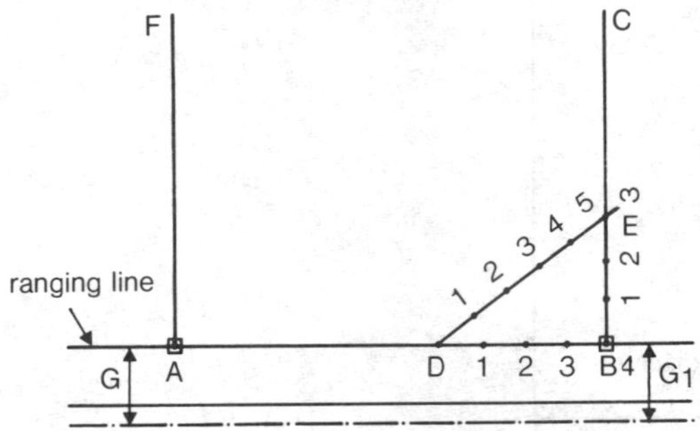

Fig. 6 The 3, 4, 5 method

The builder's square method

1 Set out the front or building line in the usual manner with pegs or marks at the required distances.
2 Place the builder's square so that the front line touches one side of the square right through its length as shown in Fig. 7.
3 Stretch a line from the corner peg so that it is parallel to the second side of the square and establish the third peg. A corner of angle 90° is thus obtained.
4 With the aid of a tape measure, mark out the length and breadth of the proposed building.
5 Transferring the builder's square to the remaining corners and repeating the above operations, a simple rectangular building can be set out.
6 After establishing the four corner pegs, profiles (separate or continuous) may be erected in the same way as described above.

Note that the builder's square can only be used with accuracy for small buildings such as farmhouses and typical single room and parlour compound buildings.

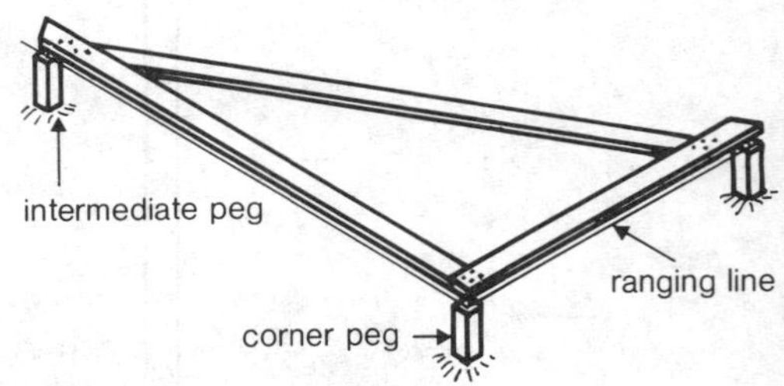

Fig. 7 The builder's square

The site square method

When a right angle is to be set out, the corner of the angle and one side of the corner are always known. One has only to find the position of the second side. The site square solves this problem in a very simple, quick and reliable way.
1 Set out the front or building line in the usual manner, with pegs or marks at the required distances as shown in Fig. 8.

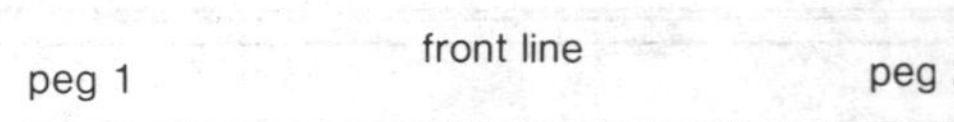

Fig. 8

2 Set up the tripod at peg 2 so that the datum rod is directly over the peg or mark which represents the corner point. Make sure that the legs are firm on the ground.
3 Release the spike screw and extend the spike so that it sits firmly on the nail or mark. Tighten the screw, as shown in Fig. 9.
4 Before mounting the instrument on to the tripod head, ensure that the locking screw is tightened. Screw on the site square. Release the locking screw. By rotating the site square, point the lower telescope along the front or building line. Tighten the locking screw.

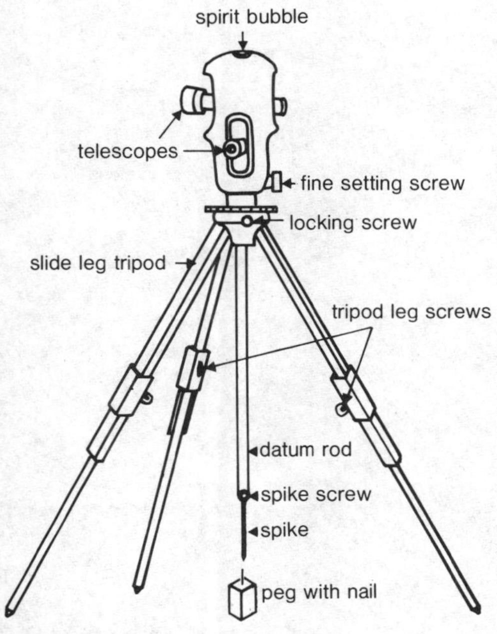

Fig. 9 The site square

5 Check the circular bubble over the top of the instrument. This will probably be found to be 'off-centre'. To correct this, release the tripod leg screws and adjust the instrument until the bubble comes into the centre of the black circle. When this is achieved, tighten the tripod leg screws. The instrument is now ready for use.
6 Sight on to peg 1 through the lower telescope to obtain the 'dead on' position by means of the fine setting screw which moves the telescope to the right or to the left and by tilting the telescope up or down as shown in Fig. 10.
7 When this position is obtained, measure the distance required to peg 3. Now by sighting through the top telescope, taking care not to rotate the instrument to the right or left, signal an assistant to move the peg sideways until it is 'dead on'. Peg 3 is now positioned at an angle of 90°.
8 By moving the site square to peg 3 and lining up on peg 2, the remaining corner peg, peg 4, can be set out using the procedures already given.
9 Profiles can be marked easily by tilting the telescope upwards, having sighted on to the peg, and placing a nail in the 'dead on' position on the profile board.

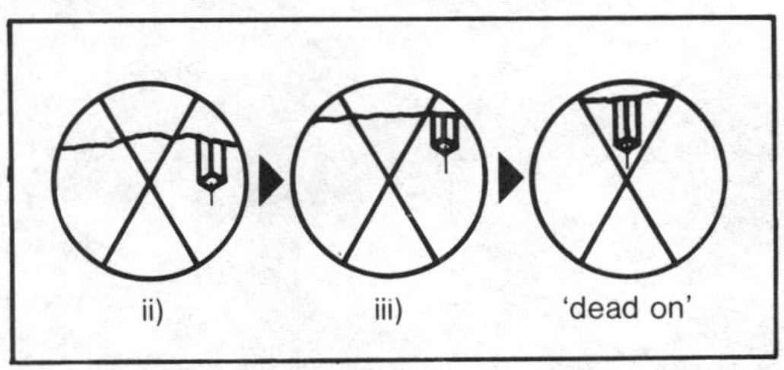

Fig. 10 Close-up view through site square

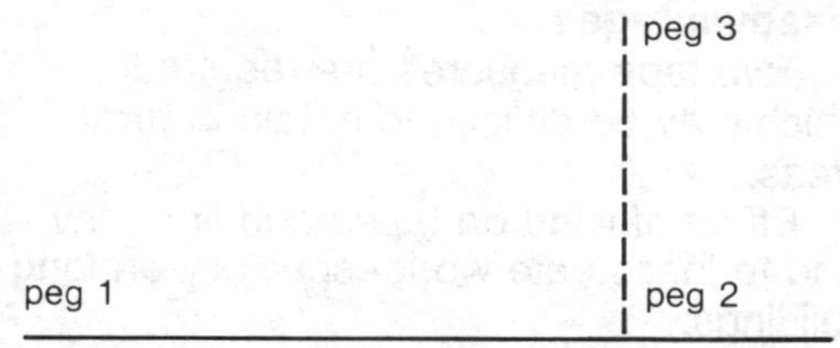

Fig. 11

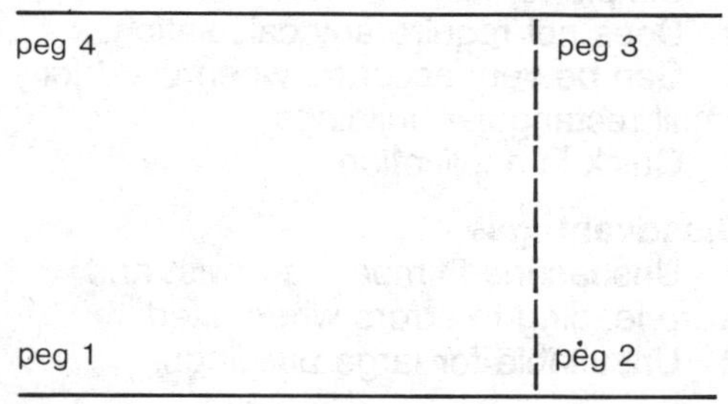

Fig. 12

Advantages and disadvantages of each method

The 3, 4, 5, method

Advantages

1 Simple to use.
2 Does not require any calculations.
3 Can be applied to large buildings.
4 Accuracy of setting out is self-evident and adjustment in the event of error is simple and quick.

Disadvantages

1 Two tape measures are required, which may be difficult to obtain in rural areas.

2 Effect of wind on tapes and line may lead to inaccurate work especially on long buildings.

The builder's square method

Advantages

1 Simple to use.

2 Does not require any calculation.

3 Can be very accurate when used for small rectangular buildings.

4 Quick in application.

Disadvantages

1 Unseasoned timber can twist and warp leading to errors when used.

2 Unsuitable for large buildings.

The site square method

Advantages

1 Can be used for large buildings with some accuracy.

2 Transfer of lines to profile boards done easily.

Disadvantages

1 The range of accuracy is limited to only 33 m.

2 Distances cannot be read off the instrument directly but need to be measured with a tape measure.

Revision questions

1 What is meant by 'site investigation'?

2 What are the advantages of site investigation?

3 Define the following terms used in setting out buildings:
a) datum peg,
b) building line,
c) ranging line,
d) builder's square.

4 Show the outline of a corner profile and mark the position of the wall thickness and the foundation concrete projections.

5 Describe briefly how the 3, 4, 5, method is used in the setting out of a small rectangular building.

6 Compare the following methods of setting out a building:
a) using the builder's square,
b) using the site square.

7 Give five reasons why it is essential to make sure that a building is correctly set out before commencing work.

4. Blocklaying and concreting tools

Blocklaying and concreting tools consist of the tools normally used by the bricklayer plus a few additional tools which have been developed to meet the specific demands of the work involved in blocklaying. These new additions are simple to make and because they are not readily available in the market, craftsmen in different areas improvise tools for their own needs resulting in different local varieties of the same tools. The same tools are called by different local names.

The blocklayer and concretor is often also a plasterer, a tiler and a drainlayer. His tool box therefore contains tools for blocklaying, plastering, tiling and drainlaying. We shall discuss some of the most important tools under the headings that follow. For the sake of convenience, we shall group the tools under the following headings:

1 Laying tools
2 Wall straightening tools
3 Cutting and trimming tools
4 Finishing tools

Laying tools

Block trowel

This is a tool which every blocklayer and concretor must possess (Fig. 13). It is used for picking up mortar from the mortar board, spreading it on the wall for bed joints, forming cross joints on blocks, cutting off excess mortar and consolidating cross joints. It is also used for tapping down laid blocks and for the rough cutting of blocks. There are two main types of blocklaying trowels i.e. the longer heavyweight and the small lightweight types. For the apprentice the latter will be found to be more suitable, while the larger type is considered more suitable for blocklaying operations.

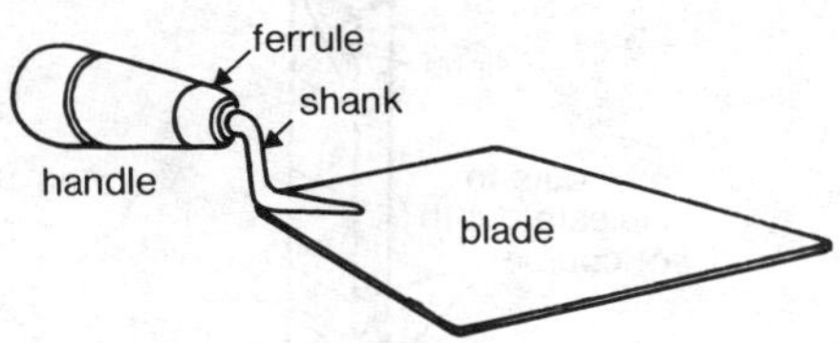

Fig. 13 Blocklaying trowel

Jointing board

The jointing board (Fig. 14) can be made quite easily from either a 25 mm thick piece of wood or a 19 mm thick piece of plywood. It measures 300 mm long by 75 mm wide, with a 25 mm gripping piece screwed to the back. It is used in practice as shown in Fig. 15.

Its use as a tool is to facilitate the filling of cross joints, to prevent mortar droppings and to facilitate the laying of mortar on the narrow webs of hollow blocks in preparation for the next course of wall. It is also used to rub the newly laid joints to a flush finish.

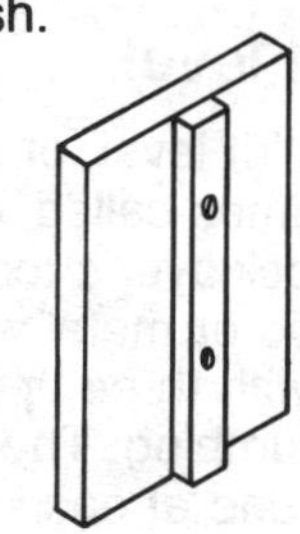

Fig. 14 Jointing board

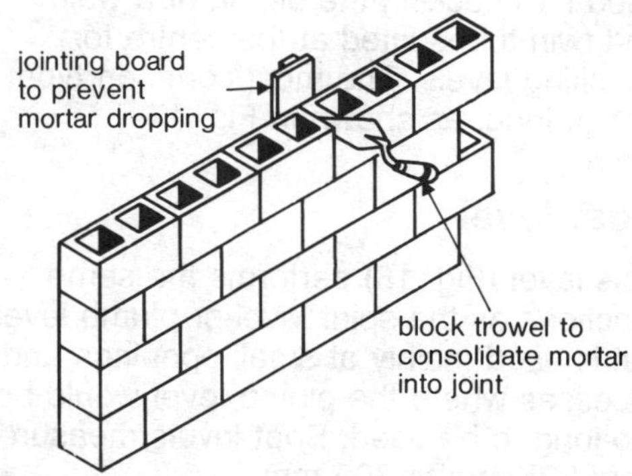

Fig. 15 Using laying tools

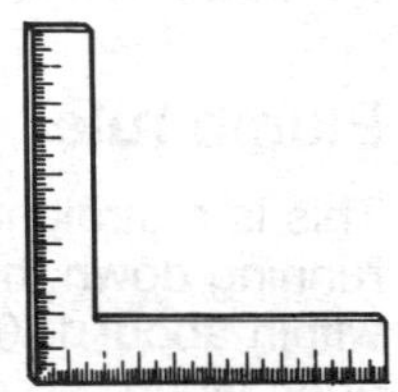

Fig. 16 Steel square

Steel square

This tool (Fig. 16) is used in checking whether or not the angle formed by two walls meeting one another is 90°. It is marked in millimetres and measures 600 mm by 450 mm.

Wall straightening tools

Spirit level

The spirit level, or plumb level as it is sometimes called, is a very useful tool in the blocklayer's tool kit. It may be made of wood or metal with straight edges. It is fitted with three spirit tubes for levelling and plumbing. These spirit tubes are fitted, one at each end with single tubes used for checking the plumb of a wall, and twin tubes fitted at the centre for checking levels. The most common type is 1 m long, as shown in Fig. 17.

Boat level

This level (Fig. 18) performs the same functions as the spirit level or plumb level, but is used mainly at small openings and recesses where the plumb level would be too long to be used. Boat levels measure from 225 mm to 300 mm.

Plumb rule

This is a straight-edge with a cut line running down the centre from the top to within about 150 mm from the bottom edge where a hole is made slightly larger

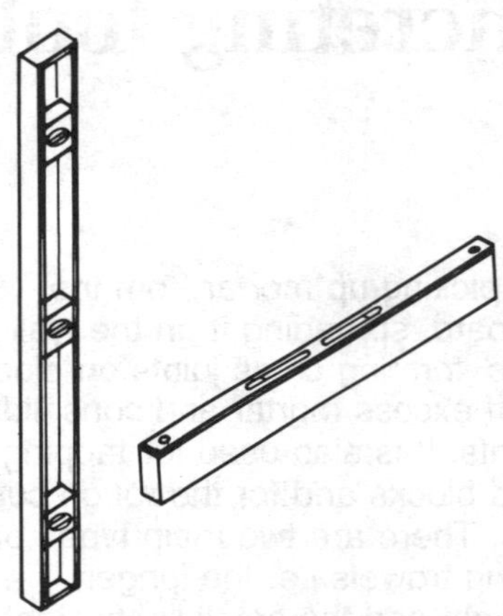
Fig. 17 Spirit levels

Fig. 19 Plumb rule

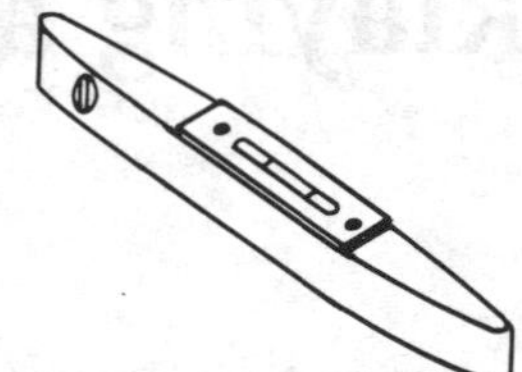
Fig. 18 Boat level

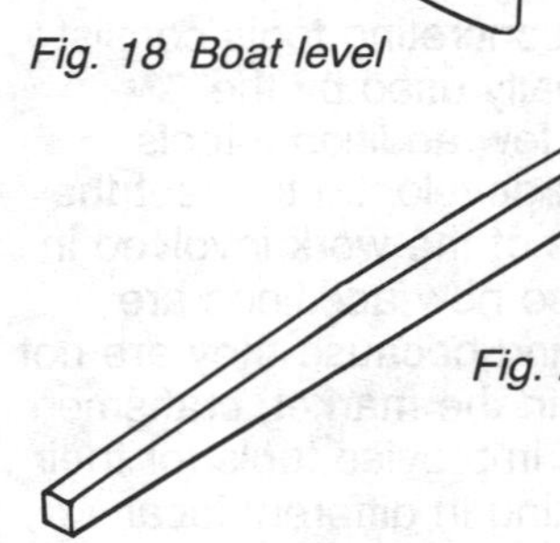
Fig. 20 Straight-edge

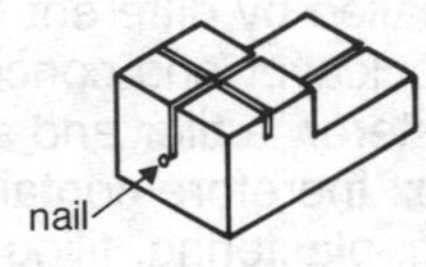

Fig. 23 Enlarged view of corner block

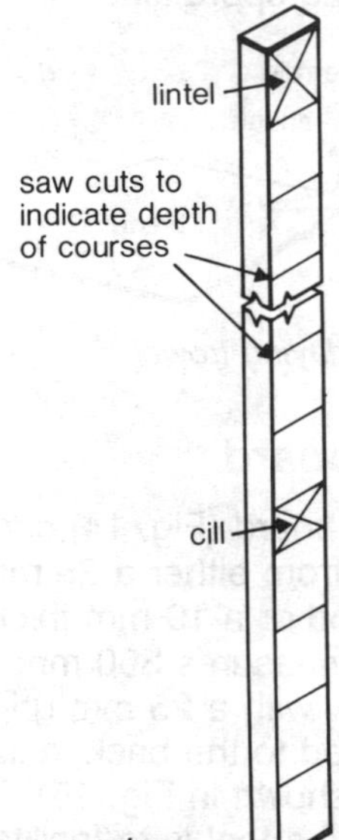

Fig. 21 Gauge rule

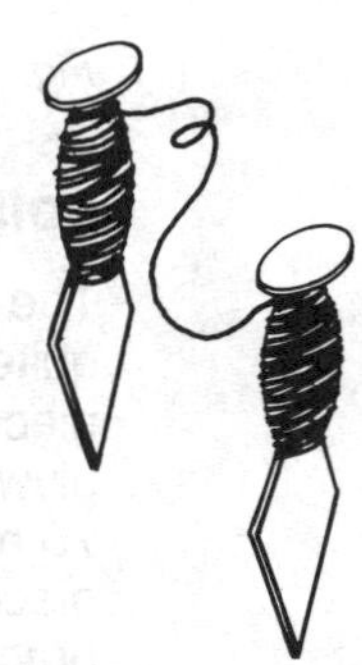
Fig. 22 Line and pins

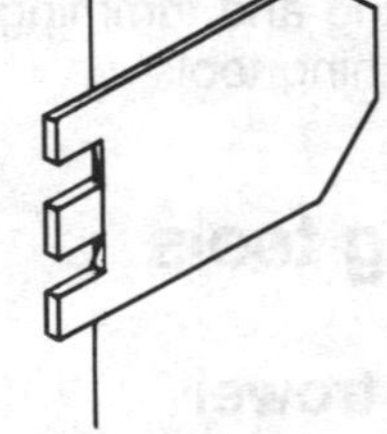
Fig. 24 Tingle plate

Fig. 25 Laying concrete blocks with the aid of a straight-edge. Note how the long, straight edge is placed on newly laid blocks and a spirit level placed on top of the straight-edge to ensure that all blocks in the course are level

than the plumb bob. The bob is an egg-shaped piece of lead held in position by a plumb line. This tool ensures a very accurate vertical check. It is shown in Fig. 19.

Straight-edge

This is a piece of timber of a length to suit its purpose, a convenient length for the blocklayer being 2 m. It is bevelled at both ends to reduce overall weight and afford convenience. The straight-edge is used to check the flatness of a newly laid piece of wall and to ensure that all blocks are laid to the same level in each course. (See Fig. 20.)

Gauge rod or rule

The gauge rule (Fig. 21) should be sufficiently long. It has the depth of block courses including the mortar joint thicknesses marked on it. In addition to the courses of blocks, various important levels, such as sill level and springing level, are marked on it.

Line and pins

The line is usually made of hemp or nylon. It is wound round the spear-like steel pins. To use this tool, the pins are each pushed into a joint at each end or quion of the wall to hold the line taut. Each course is then laid using the line as a guide, the blocks being laid to leave just a strip of daylight between their edges and the line so that they do not actually touch it. (See Fig. 22.)

Corner blocks

Where glazed work is involved and the joints are so tight that pins cannot be inserted into them, corner blocks are used instead of pins to hold the line taut. The blocklayer usually makes his own corner blocks. (See Fig. 23.)

Tingle plate

This is a small metal plate with three fingers at one end used to take up the sag of the line. A common size is 100 mm. (See Fig. 24.)

Cutting and trimming tools

Bolster

The bolster (Fig. 26) is used for the clean cutting of blocks and concrete components. It has a blade whose cutting edge ranges from 75 mm to 125 mm wide with a slight convex curve. It is hardened for cutting. The stock or striking end is left unhardened to prevent pieces of metal breaking and flying off when struck with a hammer.

Club hammer

The club hammer (Fig. 27) is made in two ranges, those with the steel head weighing 1·135 kg and those with a 1·81 kg head. It is fixed to a handle about 225 mm long. The tool is used together with the bolster for cutting purposes. (See Fig. 28.)

Skutch

The skutch (Fig. 29) consists of a stock, wedge and reversible blade. It is used for trimming a block when cut.

Square and bevel

The square and bevel (Fig. 30) are used for marking splayed cuttings.

Cold chisel

The cold chisel (Fig. 31) is made in varying lengths and thicknesses from 6·25 mm x 400 mm to 25 mm x 450 mm. They are made of steel with the cutting edge and the stock treated in a similar manner to those of a bolster.

Block axe

The block axe (Fig. 32) is very popular among blocklayers. It is used for cutting blocks into required dimensions and for cleaning the cut edges in much the same way as a scutch is used. Hollow blocks are difficult to cut with success using a

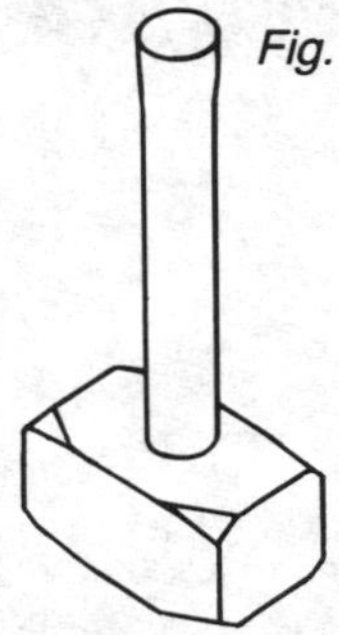

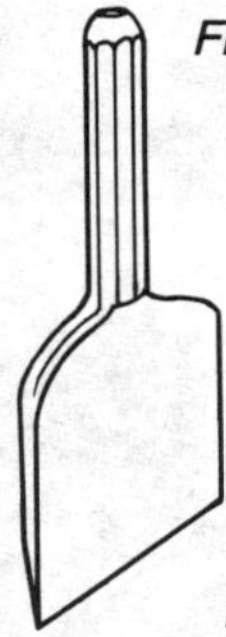

Fig. 26 Bolster

Fig. 27 Club hammer

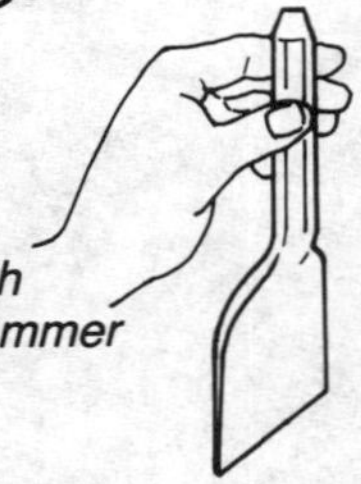

Fig. 28 Cutting with bolster and hammer

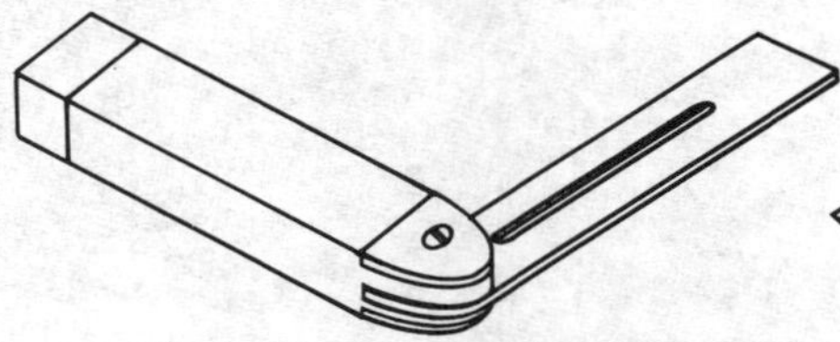

Fig. 30 Square and bevel

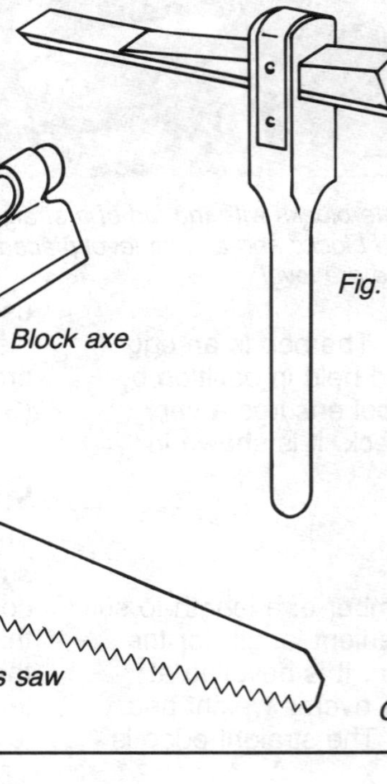

Fig. 29 Scutch

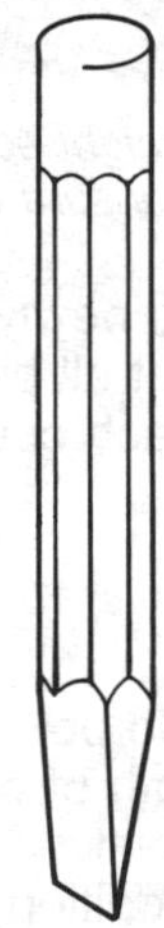

Fig. 31 Cold chisel

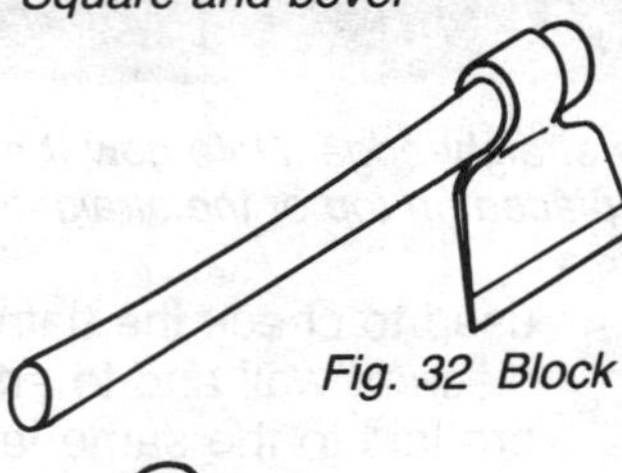

Fig. 32 Block axe

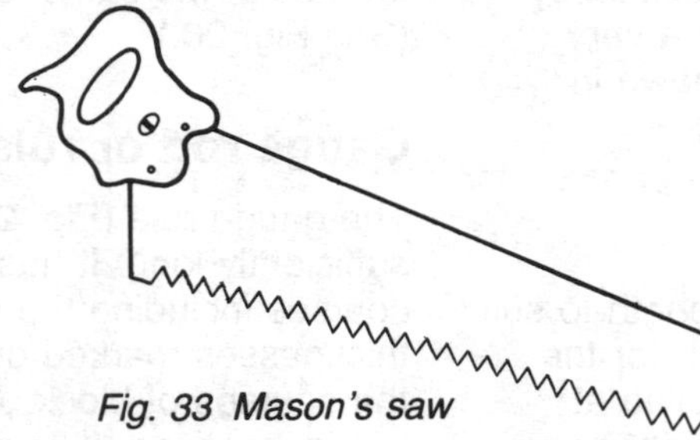

Fig. 33 Mason's saw

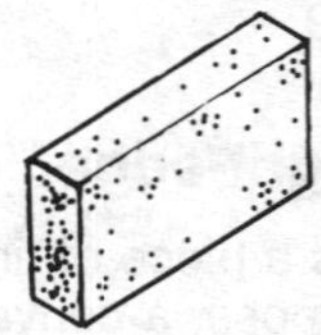

Fig. 34 Carborundum stone

bolster and hammer because of the thickness of the ribs bounding the holes. The cutting is, therefore, often done with a mason's saw.

Mason's hand saw

The hand saw (Fig. 33), with 12 mm teeth, is used for cutting soft stones and sandcrete blocks. The saw is similar in all respects to the carpenter's hand saw except that the mason's hand saw is set with chisel-shaped teeth of the necessary clearance for the saw cut. When cutting a block, with the saw, the block is first soaked with water and kept wet throughout the cutting operation. The water acts as a lubricant to minimise friction between the saw and the block.

Carborundum stone

A carborundum stone (Fig. 34) is used for rubbing down the cut edges of blocks.

Finishing tools

Pointing trowel

This is smaller in shape and lighter than the laying trowel described above. It is used for filling in the joints of a block wall during a pointing operation. (See Fig. 35.)

Hawk

The hawk (Fig. 36) is a hand board made from wood, steel or aluminium and is used to carry mortar while pointing or plastering.

Fig. 35 Pointing trowel

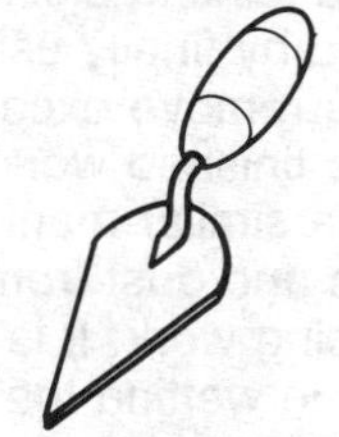

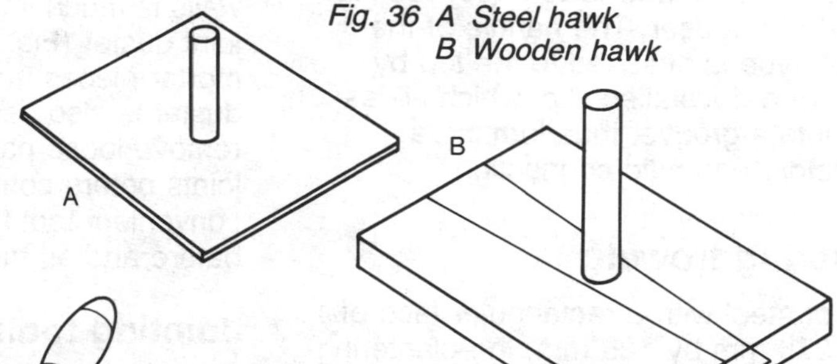

Fig. 36 A Steel hawk
B Wooden hawk

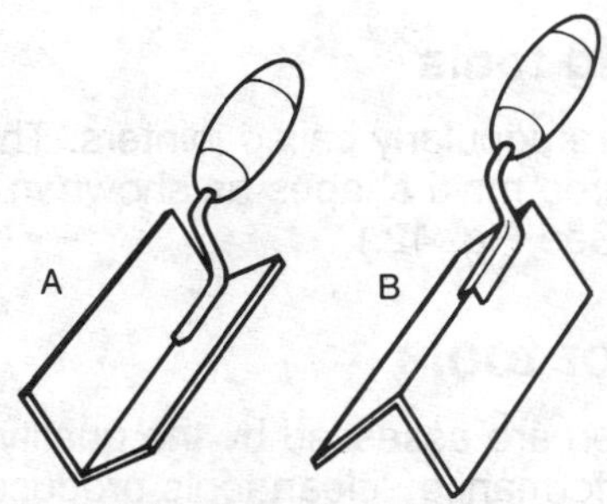

Fig. 39 A Internal angle tool
B External angle tool

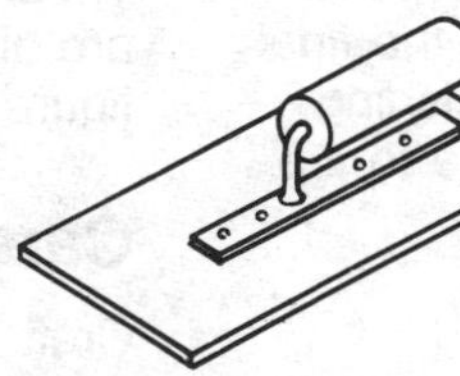

Fig. 37 Plastering trowel

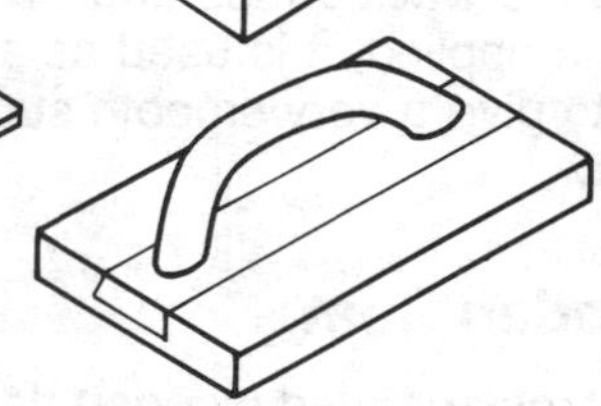

Fig. 38 Wooden float

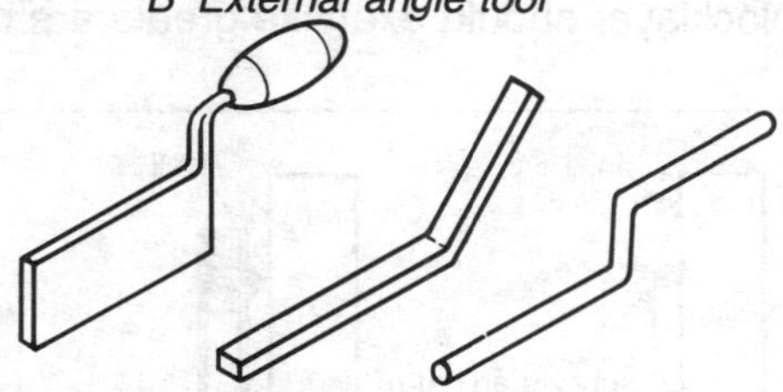

Fig. 42 Jointers

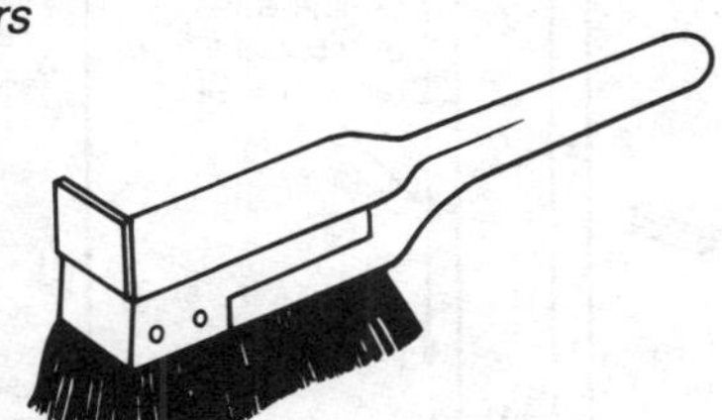

Fig. 40 Wire brush

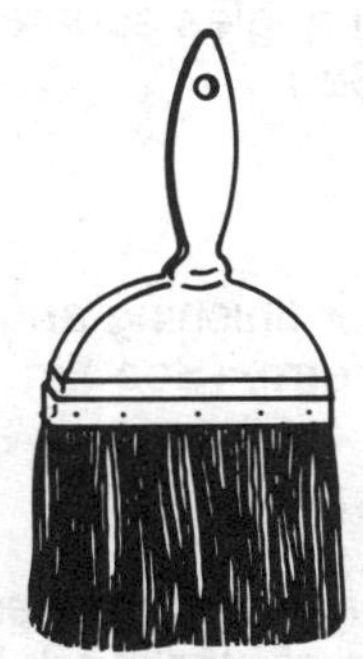

Fig. 41 Duster

The aluminium type has a wooden handle which is screwed on to a 225 mm by 225 mm plate with a foam disc in between for comfort of the user. The handle of the wooden type is attached to the top by means of a dovetailed slip, which slides easily into a groove, thus forming a satisfactory and rigid connection.

Plastering trowel

Made of steel with a rectangular face of about 275 mm by 125 mm, the plastering trowel (Fig. 37) is used for laying mortar on to the wall surface and, when the final coat is applied, it is used as a finishing tool to give a very smooth surface to the walls.

Wooden float

The cross-grained wooden float, as the name implies, has grains running across the sole. It is used for scouring finishing coats and making angles square and clean. (See Fig. 38.)

Angle tools

These are used for finishing angles in a neat and square form. (See Fig. 39.)

Wire brush

Used for miscellaneous purposes such as cleaning the joints of blockwork before pointing operations, removing dirt from plastered surfaces before a painting operation and cleaning tiled work. (See Fig. 40.)

Joint duster

The job of pointing and jointing block walls is much improved by finally using a joint duster (Fig. 41) to remove excess mortar pieces from the finished work. The duster is also used in a similar manner to remove loose particles and dust from the joints before commencing work. It is a convenient tool to use in wetting the joints before and as the work proceeds.

Jointing tools

These are popularly called jointers. They are of three main shapes as shown in the figure. (See Fig. 42.)

Care of tools

Craftsmen are assessed by the quality of their performance: 'clean tools produce a clean job' as the maxim goes. The blocklayer should exercise great care of his tools. Since he often works in the open and is subject to weather conditions such as frequent rain, tools made of metal can easily spoil through rusting. In the case of timber tools, exposure to excessive heat or wetness causes them to loose their shape. Taking good care of your tools prolongs their life in addition to making working easier and more efficient.

It should be a habit of the craftsman to clean and dry his tools after each day's work. It is advisable to set aside an hour after the close of work each Friday or Saturday for thorough cleaning, scouring and oiling of tools. During such periodic cleaning, apply linseed oil to spirit levels and plumb rules, and use rags to wipe all tools with a thin layer of oil. Dried mortar should be removed from both the handle and the blade of trowels. Check plumb levels with an established plumb line and horizontal line to make sure that they give correct readings, as shown in Fig. 43.

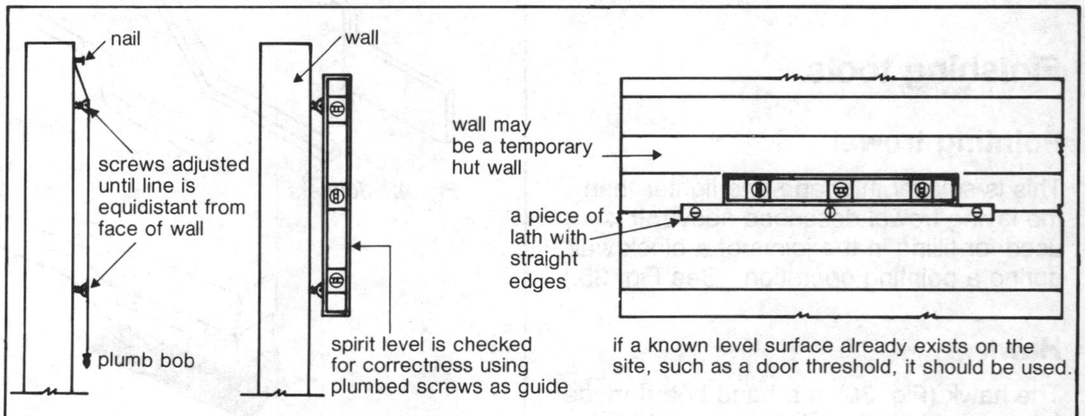

Fig. 43 Method of checking a spirit level

The reward to the owner of clean tools is great and everyone should form the habit of periodically cleaning and checking the correctness of his tools.

For concrete mixers, check the oil and water levels of the engine. Grease the hoist brushes where fitted. After the day's work, clean the baffle blades and wash out the drum. Remove mortar from the body before it dries out. Clean the machine thoroughly once a week, checking the adjustment of the drive chains, clutches, ets.

Clean buckets thoroughly after each day's work. Encrusted buckets are heavy, hold less material and can spoil mortars for plaster. Keep the cutting edge of shovels clean and sharp. This makes them lighter to handle and they will load easily. Wash off any cement at the end of the day.

Wheelbarrows should have their axles oiled daily to give easy running and also avoid excessive wear. Caked materials must never be removed with hammers. Hose out a wheelbarrow after work.

A cold chisel with a mushroom head must not be used until the skirt is ground off as metal fragments can fly off and injure the eyes.

Revision questions

1. List five tools under each of the following subheadings:
   a) blocklaying,
   b) wall straightening,
   c) cutting and marking,
   d) finishing.
2. Produce neat sketches of the following blocklaying tools:
   a) blocklaying trowel,
   b) jointing board,
   c) line and pins,
   d) gauge rule,
   e) block axe,
   f) club hammer.
3. Describe, in each case, the correct method of using the following tools:
   a) tingle plate, b) steel square,
   c) plumb bob, d) wooden float,
   e) block duster.
4. Why is it important frequently to clean and check tools for accuracy?
5. With the aid of neat sketches, show the method of checking a spirit level for accuracy.
6. Describe how the following tools should be cleaned after each working day: a) wheelbarrow , b) spirit level,
   c) bucket.

5. Walling materials

Cement

Cement is made from earthy materials such as clay, chalk and shale. When mixed with water, cement has the ability to bind particles of gravel or sand together and when set hard is capable of performing the functions expected of building and civil engineering structures.

The most common cement used for ordinary building works is called the ordinary Portland cement. Ordinary Portland cement is made to the British Standard Specification No. 12: 1978 or some other suitable specification which may exist in your country. These specifications specify the material to be used in the manufacture of the cement, the method of manufacture, methods of taking samples for testing and methods of testing for fineness, chemical composition, strength and setting time of the cement paste.

The raw materials of Portland cement are calcium carbonate in the form of chalk or limestone, alumina and silica in the form of clay.

Manufacture

Limestone or chalk is quarried, crushed and conveyed to a silo. The clay is passed through a washmill where large particles and impurities are removed, after which the clay is passed to storage tanks. The prepared clay is then mixed with the crushed limestone in the wet grinding mill in a proportion of 3½ parts of limestone to 1 part of clay measured by weight. The mixture of limestone and clay is ground in the mill to a fine state and thoroughly mixed. In this state, the mixture is known as slurry. It is further passed through a fine sieve and stored in slurry tanks, where it is stirred continually, to prevent settlement.

From the slurry tanks, the cement slurry is pumped to a rotary cement kiln, which is a long cylinder lined with heat resistant bricks known as firebricks. The kiln rests on rollers and slopes slightly. The slurry enters at the upper end of the kiln, while at the lower end a mixture of powdered coal, known as pulverised coal, and compressed air produces a very hot flame in the kiln. At the lower end the heat is about 1500 °C while the temperature at the top is about 250 °C. (See Fig. 44.)

On entering the kiln the slurry is first dried then, on passing through the central zone, the calcium carbonate content loses its carbon dioxide and becomes calcium oxide or quicklime. In the hot lower zone, the three oxides of calcium, aluminium and silicon become combined chemically to form cement clinker. The clinker is in small grey balls. After cooling, the clinker is proportioned and about 3% gypsum is added. The mixture is then ground to a very fine powder. The purpose of gypsum is to regulate the setting time of the finished cement.

Modified Portland cements

There are a number of Portland cements modified to suit the conditions that exist on site or the use to which buildings are put. It should be borne in mind that the forms of cement described below are mostly from the family of Portland cements and that, although the initial setting and hardening times may be greatly influenced by modifications, the ultimate strength gained is typical of the ordinary Portland cement already discussed.

Rapid-hardening cement

This type of cement is ordinary Portland cement ground to a finer powder. The finer the grinding, the faster is the setting time. It is useful where a faster rate of strength gain, which cannot be obtained by using the ordinary Portland grade, is necessary.

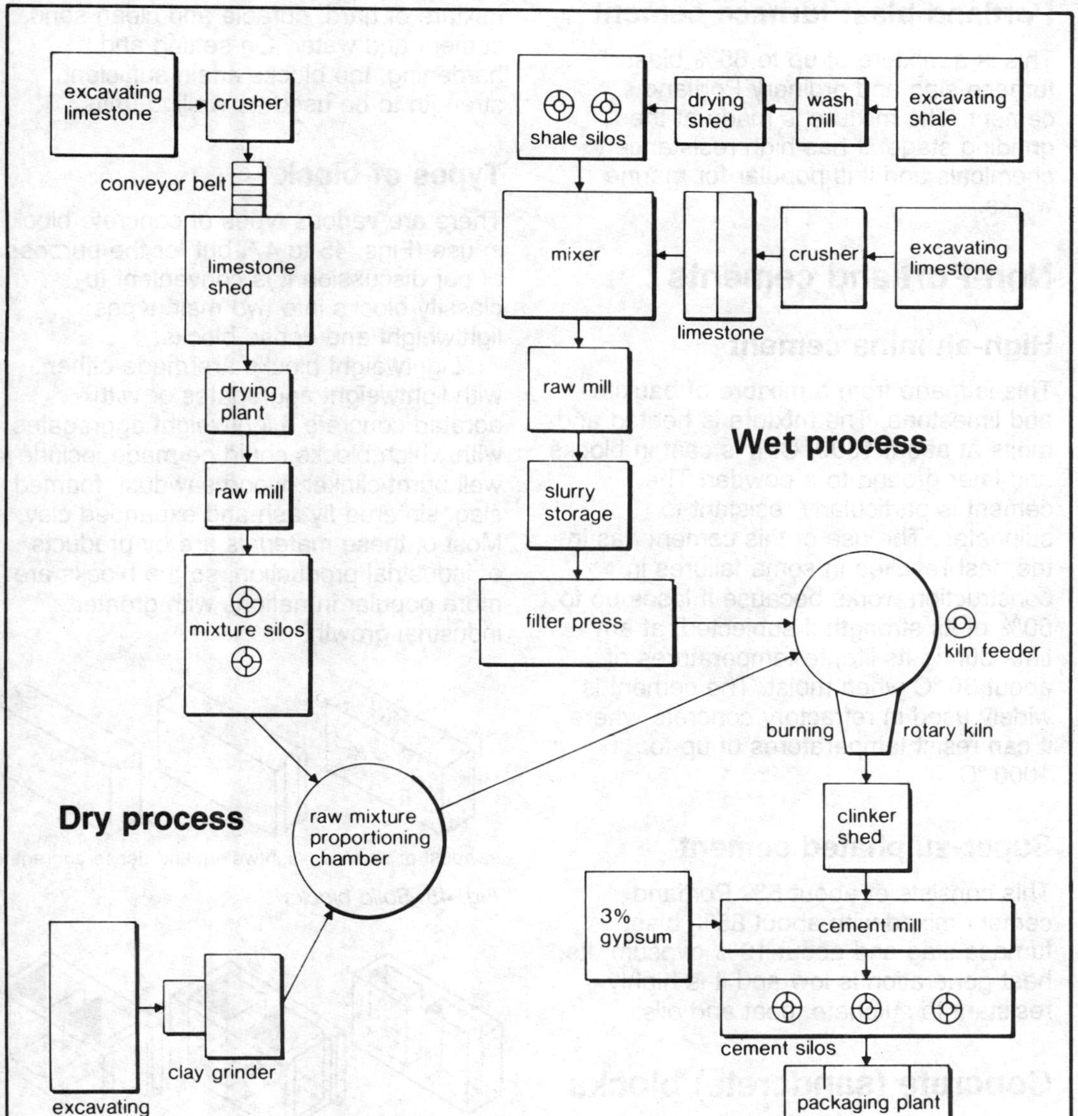

Fig. 44 Processes of cement manufacture

Extra rapid-hardening cement

This cement is particularly useful in temperate zones where winter weather conditions adversely affect construction works. It is used to get early setting and hardening of concrete.

Rapid-hardening cement can have 2% calcium chloride added to it to make this type of cement. This accelerator is either added at the time of grinding the clinker or during the mixing of the concrete. In the latter case, it must be added in liquid form.

It is not recommended for works in tropical countries because of its high rate of heat generation during setting. It is not even suitable for mass concrete dams as is often mistakenly believed by students. The heat on hydration is not easily lost from such heavy works and the stored heat is capable of cracking, shrinking or even destroying the works on cooling.

Low-heat Portland cement

This cement contains less tricalcium than ordinary Portland cement and evolves less heat during hardening. Whilst ordinary Portland cement evolves about 100 calories per gram in 28 days, low-heat cement evolves up to 70 calories per gram in 28 days. It is recommended for mass concrete work, such as dams, abutments and where trapped heat would cause damage to the concrete.

Sulphate-resisting cement

This contains less tricalcium silicate than ordinary Portland cement and therefore is more resistant to sulphate attack. Sulphates are salts found in water and in some building materials. They attack cement, making it lose its binding quality. Sulphate-attack also causes considerable damage in cement-based works such as plastering, tiling, etc.

White cement

White cement is made by using china clay (kaolin) which is free from iron oxides, instead of ordinary clay. It is the iron oxide in clay that gives the ordinary Portland cement its grey colour.

Coloured cements

Colours in cement can be obtained by blending pigments with either grey cement or white cement, depending on the colour desired. The colour pigment is either added at the cement works or by the user during mixing.

Masonry cement

This is ordinary Portland cement with plasticiser and an inert filler such as chalk or silica added. It is used mainly for blocklaying and similar works.

Water-repellent cement

During the grinding of ordinary Portland cement clinker, a quantity of metallic soap or gypsum with tannic acid is added to render the finished product water-repellent. The cement is particularly useful for water-repellent renderings and similar jobs.

Portland blast furnace cement

This is a mixture of up to 65% blast furnace slag and ordinary Portland cement. The mixture is made at the grinding stage. It has high resistance to chemicals and it is popular for marine works.

Non-Portland cements

High-alumina cement

This is made from a mixture of bauxite and limestone. The mixture is heated and melts at about 1600 °C. It is cast in blocks and later ground to a powder. The cement is particularly resistant to sulphates. The use of this cement has in the past resulted in some failures in construction works because it loses up to 50% of its strength if subjected, at any time during its life, to temperatures of about 30 °C when moist. The cement is widely used in refractory concrete where it can resist temperatures of up to 1000 °C.

Super-sulphated cement

This consists of about 5% Portland cement mixed with about 85% blast furnace slag and about 10% gypsum. Its heat generation is low and it is highly resistant to sulphate, peat and oils.

Concrete (sandcrete) blocks

Concrete blocks, sometimes called sandcrete blocks, are made from a mixture of hard, durable and clean sand, cement and water. On setting and hardening, the blocks attain sufficient strength to be used as walling units.

Types of block

There are various types of concrete block in use (Figs. 45 to 47) but for the purpose of our discussion it is convenient to classify blocks into two main types: lightweight and dense blocks.

Lightweight blocks are made either with lightweight aggregates or with aerated concrete. Lightweight aggregates with which blocks could be made include well burnt clinker, wood sawdust, foamed slag, sintered fly ash and expanded clay. Most of these materials are by-products of industrial production, so the blocks are more popular in nations with greater industrial growth.

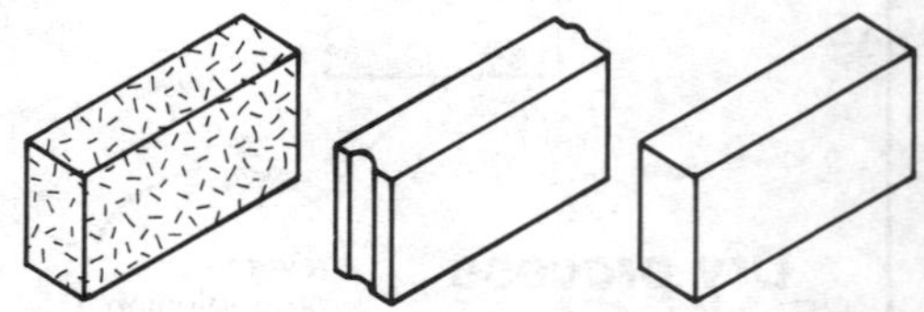

Fig. 45 Solid blocks

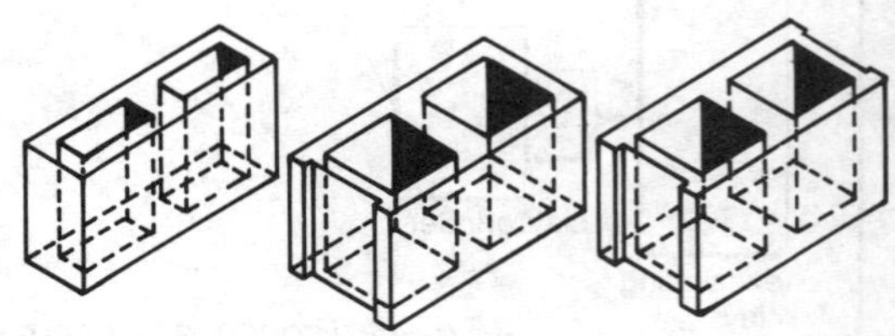

Fig. 46 Hollow blocks

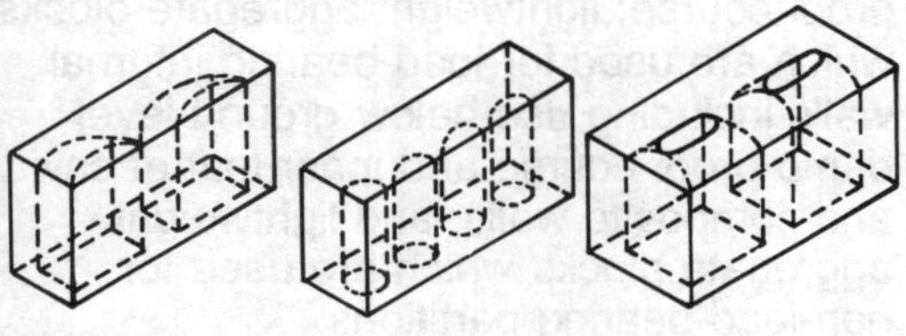

Fig. 47 Cellular blocks

Aerated concrete block is defined in B.S. 6073: Parts 1 and 2: 1981 as consisting essentially of an inorganic cementing agent with or without the addition of suitable fine inorganic aggregates. The aerated structure is formed either by generation of a gas by a chemical action within the mix prior to hardening or by mechanical incorporation of air or other gas into the mix with the aid of suitable chemical foaming agents and mixing devices.

Dense blocks are made from dense aggregates as specified in B.S. 882: 1983. They include natural sand and crushed rock. Blocks made from such aggregates normally have densities of 1920 to 2080 kg/m^3. Dense blocks may be solid or hollow. Hollow blocks have a relatively small bearing area and exert up to four times as much pressure on the mortar as lightweight blocks or burnt clay bricks. Saturated dense blocks are liable to 'swim' on the mortar, so the mortar used has to be fairly stiff to avoid subsequent settlement, which leads to cracks in the wall.

Type of block	Material used	Surface or features
Clinker	By-products of high-temperature coal boiler furnaces	Angular in shape: dark in colour
Foamed slag	By-products from iron manufacture. Blast furnace slag is chilled and expanded then crushed and screened	Angular and porous
Sintered fly ash	By-product from power stations using pulverised coal fuel	Porous nodules and pellets both fine and coarse graded
Wood sawdust	By-product of wood working shops. Mixed with sand and cement and pressed into solid blocks	Porous. Wood shavings seen on surface
Expanded clay	Clay expanded with sulphite ley and burnt	Angular or rounded depending on method of manufacture
Sandcrete	Coarse natural sand or crushed rock dust mixed with cement and water. Pressed into shape	Rough porous surface. Cement grey colour
Aerated concrete	Siliceous material with addition of aluminium powder	Porous

Table 1 Summary of types of blocks and their characteristics

Type of blocks	Actual size	Nominal size
	Length, Height Width (mm)	Length, Height Width (mm)
Dense aggregate	440 × 140 × 90 440 × 215 × 215 440 × 215 × 140	450 × 150 × 100 450 × 225 × 225 450 × 225 × 150
Lightweight aggregate blocks. Load-bearing	440 × 140 × 90 440 × 215 × 140 140 × 150 × 215 440 × 140 × 65	450 × 150 × 100 450 × 225 × 150 150 × 150 × 225 450 × 150 × 75
Non-load-bearing lightweight blocks	450 × 215 × 40 440 × 215 × 65	450 × 225 × 50 450 × 225 × 75

Table 2 Sizes of blocks in common use

Size of blocks

Building blocks come in a variety of sizes, some being made to suit the owner's handling convenience. Such blocks are non-standard as they are not factory produced. Many of the blocks produced in developing countries are in this category. Commercial blocks are produced to meet standard sizes. Blocks in this category are made in accordance with B.S. 6073: Parts 1 and 2: 1981 or a similar standard.

Table 2 gives the dimensions of blocks incorporated in the latest amendment to B.S. 6073: Parts and 2 published by courtesy of B.S.I. The actual dimensions are without joints while the nominal dimensions allow for the thickness of joints.

A great number of blocks are wasted during the cutting of halves and other required pieces. To avoid this wastage, the manufacturers should be consulted so that halves, three-quarters etc. as would be required on the job are made by the blockworks. Such consultation is also necessary at the design stage so that the layout of the walls can as far as possible make use of full and half length units. Thus economy in construction is achieved.

Classification of blocks

Building blocks are classified according to specified properties. The manufacturers usually have good varieties to choose from depending on the requirements of the user.

There are basically three grades of building blocks. These are: dense aggregate blocks which are suitable for general use in load-bearing external walls including works below ground level damp proof course; lightweight aggregate blocks which are used for load-bearing internal walls including use below ground level damp proof course and inner leaf of cavity and composite walls; and lightweight aggregate blocks which are used for non-load-bearing partitions.

Materials used for making building blocks include natural sand and crushed stone, by-products of industrial production such as clinker, foamed or air-cooled slag, sintered pulverised fuel ash, sawdust and wood shavings and expanded clay or shale. Among these various materials, natural sand and crushed stone are more likely to be available in most developing countries. Frequently, choice of aggregate is determined by what is locally available at an economic rate.

In most countries, classifications are defined according to specified properties and uses without reference to materials or method of manufacture, provided that the resulting blocks comply with such specifications. The choice of material is left to the block manufacturer, as it is the selection of the correct ingredients and proper control over quality, density, shape and dimensional accuracy that will ultimately determine the class to which a finished product belongs.

The following information will serve as a guide in determining the grade of a block. First grade blocks are those with not less than 1500 kg/m^3 density and a minimum average compressive strength of 7·0 N/mm^2; second grade blocks are those with a block density less than 1500 kg/m^3 and a minimum average compressive strength of 2·8 N/mm^2 but more than 625 kg/m^3 and a minimum average compressive strength of 7·0 N/mm^2; third grade blocks are those with a block density of less than 1500 kg/m^3.

Blocks are extensively used as facing works in many developing countries. Apart from the traditional sand-textured flat surfaced blocks, there exist nowadays a wide range of specially designed facing blocks manufactured with profiled surfaces or with textured faces such as those using exposed aggregates.

Properties of blocks

Every walling material undergoes dimensional changes with changes in temperature, especially if accompanied by drying and wetting. These kinds of changes can cause cracking of block walls unless adequate precautions are taken. The changes may be small but, with concrete or sandcrete blocks, the problem may be aggravated due to additional non-reversible shrinkage caused by chemical changes associated mainly with the processes of carbonation, hydration and curing.

Strength

Many purchasers of blocks do not bother to check the strength of their blocks before building them into walls. Architects may specify the required strength to comply with established standards but often compliance with specifications is taken for granted. Failure in this regard may be the cause of the many ugly cracks that appear on blocks walls. In this context, strength is the average compressive strength of a test sample of ten blocks, and the compressive strength of the weakest individual block in the test sample must not be less than 80% of the average value. In this way control is exercised on the quality of the blocks in the consignment as a whole. This test is only relevant to load-bearing blocks.

Non-load-bearing blocks are subjected to a transverse breaking test load and the result must meet specified standards. Solid, hollow and cellular load-bearing blocks are tested for compressive strength as described above. To obtain the compressive strength of the block, the crushing load is divided by the cross-sectional area. Thus there is no difference in the load carrying capacity of solid blocks and of hollow or cellular blocks of equal compressive strength.

The design requirements for concrete or sandcrete blockwork are covered in C.P. 111: Part 2: 1970 with Amendment No. 1. The reader should refer to this document for further details.

Fire resistance

A concrete block is an excellent fire resistant material for wall construction. From the records of fire incidents, the performance of solid load-bearing blockwork is quite high and there does not seem to be any problem. The Building Regulations 1985 calls for an unplastered thickness of just 100 mm for one hour's fire resistance with a wide range of materials including blockwork.

Durability

Unlike brickwork, most blockworks are treated with a covering finish, such as rendering and painting, tyrolean, tiling, etc. These finishes serve to protect the blockwork from direct contact with adverse weather conditions. In industrial areas atmospheric pollution may be particularly strong affecting the durability of facing blocks. In the type of weather conditions that prevail in most tropical areas, no problems should be encountered with the durability of concrete blockwork.

Thermal insulation

In the tropics the best building materials are those which are capable of excluding heat from the building. Solid blockwork of dense blocks does not serve this purpose as well as hollow and cellular blockwork. Lightweight concrete blocks provide good thermal insulation. In general, the lighter and more porous the block, the better will be its insulating value. Therefore holes in its insulating value. Therefore holes in hollow blocks should not be filled during building operations.

Sound transmission and absorption

Facing blocks with an open texture may be unsuitable as a barrier to sound transmission unless at least one side is properly plastered or painted to seal the cavities in the texture. The problem of sound transmission is made worse where continuous holes are made in walls for electrical ducts, outlets, etc. Lightweight aggregate blocks and open textured blocks perform better than dense concrete blocks in sound absorption. The sound absorptive qualities of these blocks are said to be moderately high particularly at high frequencies.

Manufacture of blocks

The various materials that can be used as aggregates for block making have already been mentioned. The choice of material will depend upon the type of block to be manufactured and it is essential to make the right choice. The most commonly used aggregate for this type of block consists of hard, durable, clean sand, mixed with cement.

Batching of materials

The mix proportions are determined carefully in order that the resultant blocks meet specific standards. Blocks for load-bearing walls should have a mix of 1 part of ordinary Portland cement to 6 parts of clean sand (1:6). Commercial producers may use proportions much weaker than this. It is therefore advisable to take all necessary precautions to ensure that blocks satisfy specified standards before buying them. A mixture of 1:4½ is used where the thickness of the web of hollow blocks is 25 mm or less. Hollow blocks for load-bearing walls should contain solid materials not less than one-half of their gross overall volume. No web should be less than 12·6 mm. To ensure uniformity of batches a gauge box ot other container whose capacity has been determined in advance such as a headpan should be used. A gauge box is shown in Fig. 48 and a headpan in Fig. 49.

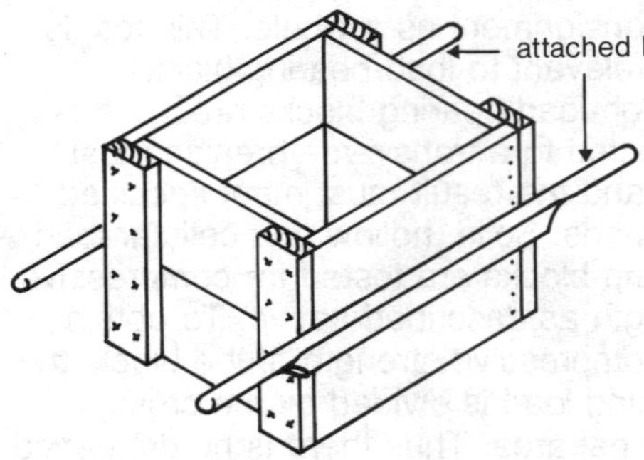

Fig. 48 Gauge box

Fig. 49 Head pan

Mixing

Where a large number of blocks is to be manufactured, the mixing should be done by mechanical means. A concrete mixer is often used, but a pan-type mixer should be preferred where it is available, as a better result is obtained. The mixing should continue until the cement content is evenly mixed with the sand. It is advisable to mix dry first and then add water in sprays. Only sufficient water to enable the hydration of the cement should be added. The mixture should cake when a quantity is pressed in between the palms without bringing out water.

Excessive water in the mix causes shrinkage and distortion of the blocks on drying. In a similar way to the measuring of the aggregates and matrix, the correct amount of water per batch should be ascertained in advance. A concrete mixer is shown in Fig. 50.

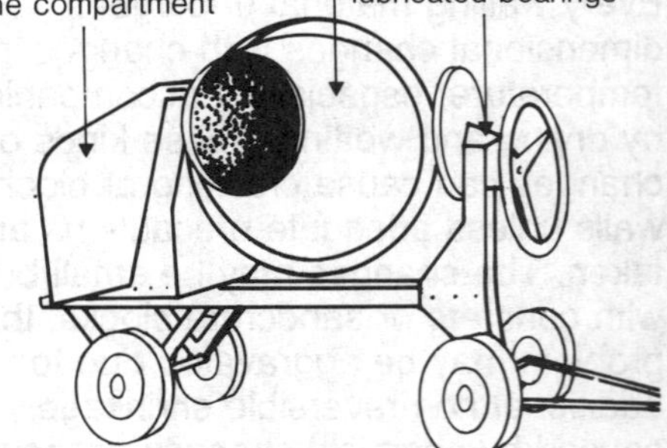

Fig. 50 Concrete mixer

Moulding

Block-moulding machines are invariably used for this purpose. The simplest of these machines are those having a weighted press which is used to tamp the blocks. The more sophisticated machines effect compaction by vibration combined with compression. Demoulding the finished products is performed by means of a guided mechanical device. Handling of freshly made blocks should be reduced to the very minimum in order to avoid distortion. To this end too, the whole casting shed and the standing for the moulding equipment itself should be so designed as to obviate any transmission of jolts or vibrations from the moulding machine to freshly made blocks. Pallet racks should always be used to handle green blocks.

Fig. 51 Concrete blocks stacked on the site according to sizes

Curing

Adequate curing of blocks is essential to ensure sufficient hydration of the cement. The method of curing employed can significantly affect the properties of the blocks. Being concrete products, strength can only be gained if blocks are subjected to conditions in which moisture is retained long enough for the setting of the cement. After the initial setting of blocks they should be kept moist be wetting until sufficient strength is gained. With ordinary Portland cement this occurs after 28 days. Because of the dangers of hot, dry weather or heavy downpours of rain, freshly made blocks should not be left in the open to dry, but always under a covered space.

Storage

Blocks should be stacked on a dry surface, away from where they will be broken by other activities in the workshop or on the site. If during the rains they are likely to be soiled by mud or buried in the wet ground, they must be stacked on a raised platform of old blocks. Stack blocks as they will be used in the wall and on no account should a stack be so high as to damage the bottom layers of blocks. All blocks should be handled with care to avoid chipping the edges and corners. This is especially important if the wall is to be left unrendered. Chipped blocks make unsightly joints so they will probably have to be discarded.

Mortar for blocklaying

The amount of cement in the mortar mix is adjusted by the blocklayer to suit the weight and suction of the blocks. If building blocks are dry and dusty for most periods of the year, high suction results. The mortar should therefore be fattened by reducing the sand content or by the addition of plasticiser.

Blocks must be kept damp by wetting before being laid in the wall. Because of the high temperature during the day, mortar should be mixed in such a quantity that it will be used up within 1½ hours of mixing. Mortar that has already hardened should not be knocked up and remixed for use, as the initial strength will already have been lost.

Binding agents

One of the most common binding agents used is ordinary Portland cement. Sometimes either lime or plasticiser is added to increase the plasticity, especially when the sand used is coarsely graded. Although some authors would like us to believe that cement/sand mortars are weak, this has not always proved to be true. Cement used for mortar should not contain lumps caused by air-setting. Bagged cement should therefore be thoroughly checked before accepting a consignment.

Sand

The sand should be hard, durable and clean. Loamy or fine sand is difficult to work with when mixed with cement. Its stickiness with the trowel slows down the pace at which the blocklayer works. On the other hand dense blocks tend to 'ride' on a mortar made with coarsely graded sand. It is also difficult to spread on the narrow webs of hollow blocks, resulting in wastage of mortar during the process of laying the blocks. Reference should be made to B.S. 1200 with regard to the quality of sand.

Mixing

Mortar is best mixed in mechanical mixers. There should be only sufficient water to ensure an acceptable state of workability. The blocklayer who is to use it should offer the best advice with regard to workability of mortar.

	Recommended proportions (by volume)		
	Cement: Lime: Sand	Cement: Sand	Cement: Sand (with plasticiser)
External walls. Walls. Work below ground. Work below damp proof course	1: 1: 6	1: 4–5	1: 5–6
Internal walls. Inner leaves of cavity walls. Backing to external solid walls. Rendered parapet wall	1: 2: 9	1: 6	1: 7
Retaining walls and the like	1: ½: 4	1: 3	1: 3

Table 3 Recommended mortar mixes for concrete or sandcrete blockwork

Lime mortar

Lime mixed with sand may be used as the binding agent for mortar. Lime was in extensive use by builders especially in temperate climates before the manufacture of cement in 1786 by Joseph Aspin. Lime requires a continuous state of moisture to gain and maintain full strength. This condition makes the use of lime unsuitable in a tropical climate which is warm and often dry.

This reason alone does not, however, preclude the use of lime for construction purposes. Limes are of two main types: 'hydraulic' and 'non-hydraulic'. The former is capable of setting under water, an important factor in the building of thick walls in water. The non-hydraulic lime needs the effect of the atmosphere to achieve its final set. Where refractory cement, which is used exclusively in the construction of fire-resisting walls, is not readily available, lime mixed with crushed burnt clay is a better substitute than cement mortar.

To prepare lime mortar, gauge the required amounts of sand and lime on a mixing stage made of cemented ground or constructed of suitable boards. With the aid of a shovel, turn the dry materials at least three times. Form a chasen and add water; turn the mixture a further three times.

Mortar plasticisers

The use of air-entraining-type plasticisers is highly recommended for blocklaying under hot, dry climatic conditions because of their advantages. There are, these days, many brands on the market capable of entraining air bubbles in the mortar. Manufacturers normally submit evidence to show that the material is suitable for the intended purpose. The manufacturer's instructions about the mixes and use should be strictly followed as failure to do so may produce poor results. The use of too much plasticiser or prolonged mixing of the mortar should be avoided, as this leads to excessive air entrainment and consequent weakening of the mortar. Roller-type mortar mills are unsuitable for the mixing process as they fail to entrain sufficient air.

Despite the fact that there is comparatively little independent information about plasticisers, many builders have made their use an established practice. Plasticisers obviate the need to use lime in the mortar and may result in economies in both capital and labour costs.

Mortar plasticisers are effective either because they entrain extra air in the mortar, or because they increase the proportion of very fine material.

There are two main effects of air entrapped in the mortar.

1 Plasticity and workability are improved allowing the water content of the mortar to be reduced by up to 50%. This automatically reduces the drying shrinkage of mortar and improves the bond with the blocks. Mortar plasticisers entrain within the mortar mix a considerable number of microscopic and disconnected bubbles of air which act as a frictionless aggregate that can be likened to 'ball bearings' of air. This makes it easier to trowel and prevents mortar adhering to the trowel, thus speeding up the rate of laying.

2 By virtue of the disconnected air bubbles in the mortar it is possible that the resistance of the mortar to rain water penetration is increased by a break in the capillary channels.

Some useful hints

Fine aggregate should be free of harmful substances and comply with the requirements set out for natural sands in B.S. 1200. High proportions of sand in the mortar will reduce its drying shrinkage but may produce a harsh and unworkable mix.

Water must be drinkable and where the supply is of doubtful quality, reference should be made to B.S. 3148. During dry weather, mortar should be used 45 minutes to one hour after mixing; in wet conditions with cooler temperatures, mixed mortar should be used up within two hours of mixing.

Bricks

Bricks are made of either clay or a mixture of sand and cement. The former is widely used for walling while the latter is used as substitute in areas where clay bricks are not in common use. In some countries,

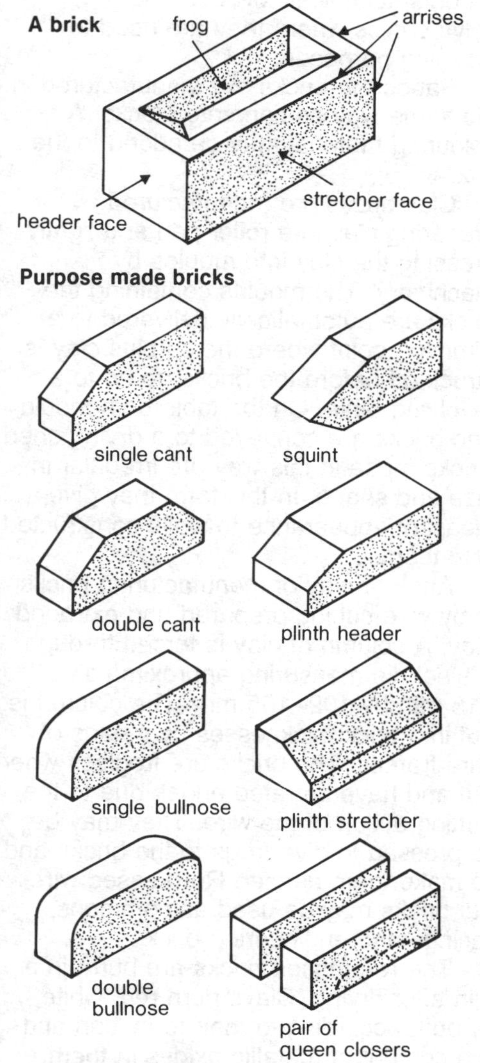

Fig. 52 Building clay bricks

sandcrete bricks are found in trade schools, technical colleges and polytechnics where they are used for teaching purposes.

Sandcrete bricks are manufactured in the same way as concrete blocks. A colouring material may be added to the mix.

Clay bricks are manufactured by preparing clay in a roller pan and lightly pressing the clay into moulds by machinery. The moulds containing the bricks are automatically delivered to a trimming point where the surplus clay is struck off before the bricks go on to a revolving table. On the table both mould and bricks are conveyed to a drying shed. Bricks made in this way are irregular in size and shape. In this form they give a pleasant appearance to walls constructed with them.

Another way of manufacturing bricks is by wire-cutting prepared and extruded clay. A column of clay is forced through a steel die measuring approximately 215 mm by 102–105 mm. The column is cut into brick thicknesses by means of wire frames. The bricks are frogless when cut and have serrated edges due to the cutting action of the wire. They may be re-pressed to give frogs to the bricks and to make them denser. Re-pressed wire-cut bricks may be used as commons, facings and engineering bricks.

The re-pressed bricks are burnt in a kiln after drying. Clays burn red, white, or buff, according to their formation and the content of metallic oxides in them, which may be up to 10%.

Revision questions

1 Describe a method of manufacturing ordinary Portland cement.

2 Cement may be manufactured using either the dry or wet systems.
a) Explain the main difference between the two systems.
b) Name the cement manufacturing works in your country and indicate in each case the system of manufacture used.

3 State in broad terms the specifications as contained in B.S. No. 12, with regard to ordinary Portland cement.

4 Describe the method of manufacturing sandcrete blocks.

5 Describe three ways by which a lightweight block can be made.

6 Make neat isometric sketches of the following building blocks:
a) solid lightweight,
b) hollow,
c) cellular.

7 What are the precautions to be taken to prevent cracking and twisting of newly made blocks?

8 Describe how a fresh load of building blocks should be stored on a building site.

9 State four essential qualities of cement mortar.

10 a) What is the purpose of adding lime to cement mortar?
b) How does plasticiser aid workability in mortar?
c) What are the main disadvantages of using smooth sand for mortar?

11 Describe how a cement/sand/lime mortar is prepared.

12 Describe the method of manufacturing wire-cut pressed bricks.

13 Make neat isometric sketches of the following types of brick:
a) plinth header,
b) double cant,
c) single bullnose,
d) a pair of king closers.

14 Produce an isometric sketch of a brick and show the following parts:
a) header face,
b) stretcher face,
c) frog,
d) arrises.

6. Bonding

Walls built with stones, blocks and bricks consist of individual units. For such walls to fulfil the functions expected of them, i.e. to be strong and durable, the unit blocks or stones must be arranged in a pre-determined pattern which must present a pleasant surface appearance and at the same time eliminate, as far as possible, straight joints. This is termed **bonding**.

Load-bearing walls must always be bonded. Several bonds are in use which give quite pleasant elevational appearances. Bonding is not only applied to the individual units in the course, but the entire course must also bond with other courses in the wall. Often there is more than one solution to a bonding problem. The craftsman will select the one which is most appropriate. The principle of bonding is illustrated in Fig. 53.

In order to solve bonding problems, there are certain rules which should be applied. It must be noted that these rules are only a guide, and that other considerations such as minimisation of wasteful cutting of units and their economical use should always be borne in mind.

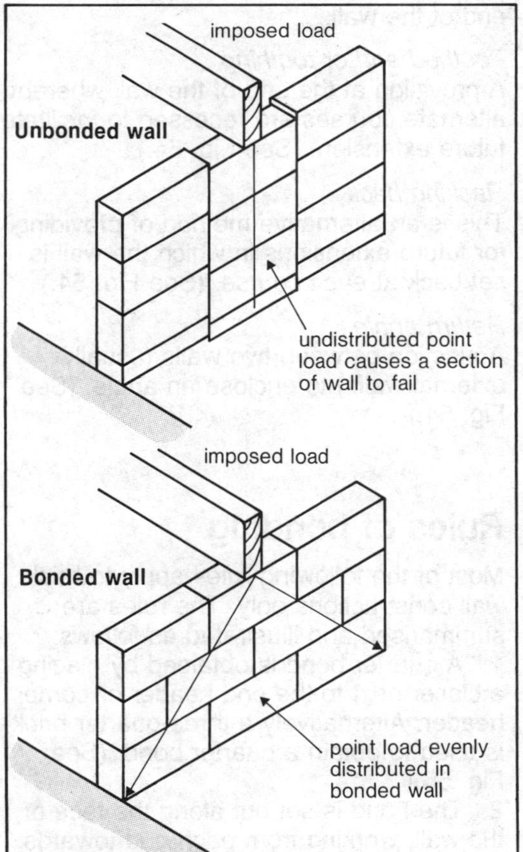

Fig. 53 The principle of bonding

Terms used in bonding

Brick
A building unit of actual size 215 mm by 102·5 mm by 65 mm. If a 10 mm mortar joint is assumed, this size becomes 225 mm by 112·5 mm by 75 mm, thus giving a gauge of four courses to 300 mm. (See Fig. 54.)

Stretcher
The face of the brick that measures 215 mm by 65 mm. (See Fig. 52.)

Header
The end of the brick measuring 102·5 mm by 65 mm. (See Fig. 52.)

Frog
The depression formed in the top of the brick during manufacture. (See Fig. 52.)

Arris
The sharp edge of the brick. (See Fig. 52.)

Queen closers
These are obtained by splitting the brick equally along its length. (See Fig. 52.)

Half bat
This is obtained by cutting a brick through the centre line across its width. (See Fig. 55.)

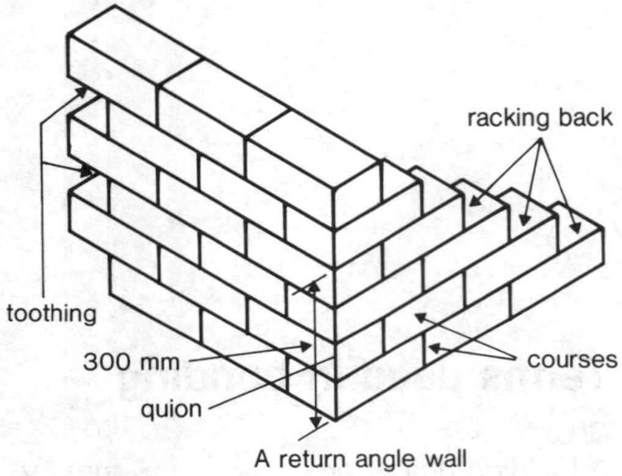

Fig. 54 Terms used in bonding

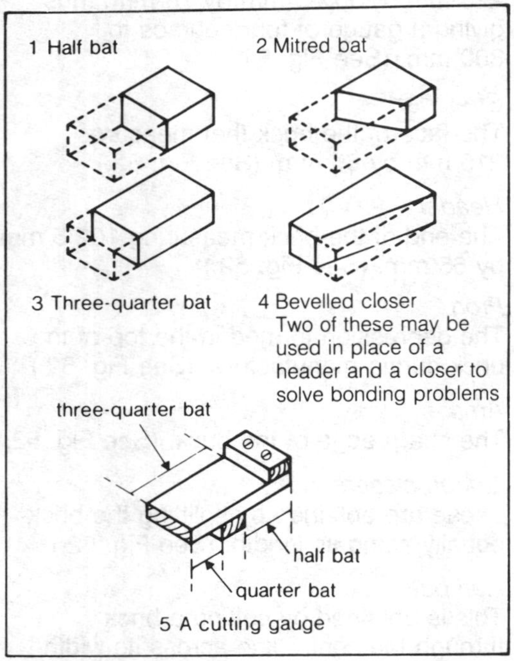

Fig. 55 Terms used in bonding

Three-quarter bat
This is obtained by cutting a closer off the bricks. (See Fig. 55.)

Course
A term applied to each row of bricks in a wall. (See Fig. 54.)

Stopped end
A plain vertical surface which forms the end of the wall.

Toothed end or toothing
A provision at the end of the wall whereby alternate courses are recessed to facilitate future extension. (See Fig. 54.)

Racking back
This is an alternative method of providing for future extensions in which the wall is set back at each course. (See Fig. 54.)

Return angle
A junction between two walls (usually external walls) to enclose an angle. (See Fig. 54.)

Rules of bonding

Most of the following rules apply to brick wall constructions only. The rules are summarised and illustrated as follows.

1 A quarter bond is obtained by placing a closer next to the end header or corner header. Alternatively a three-quarter brick is used to obtain a quarter bond. (See Fig. 56.)

2 The bond is set out along the face of the wall, working from each end towards the centre, the end bricks in each course corresponding on the face. (See Fig. 57.)

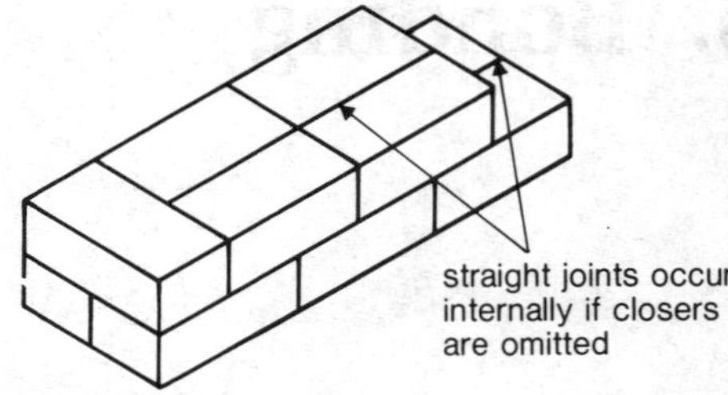

Fig. 56 Rules of bonding (rule 1)

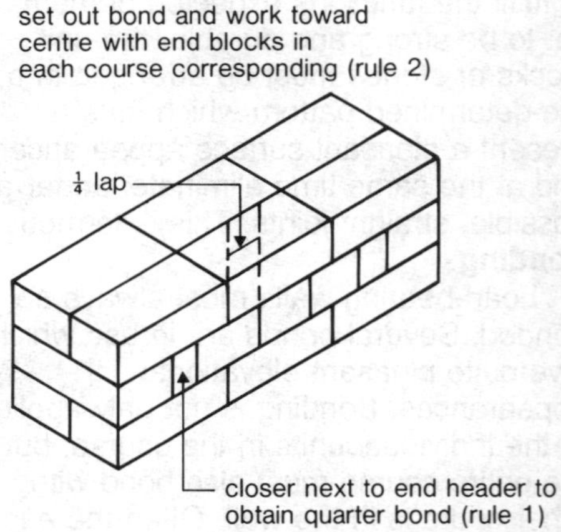

Fig. 57 Rules of bonding (rules 1 and 2)

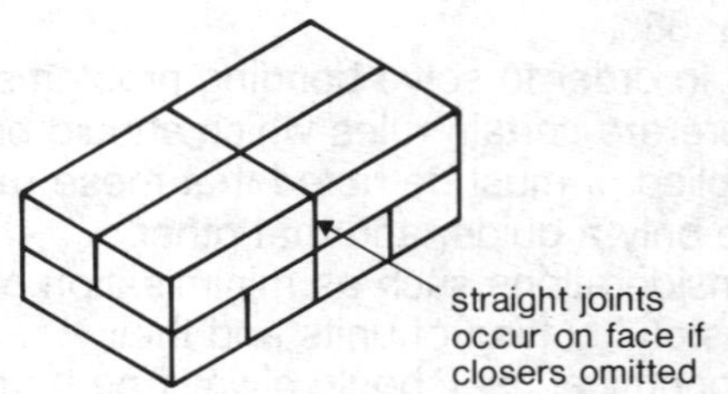

Fig. 58 Rules of bonding (rule 1)

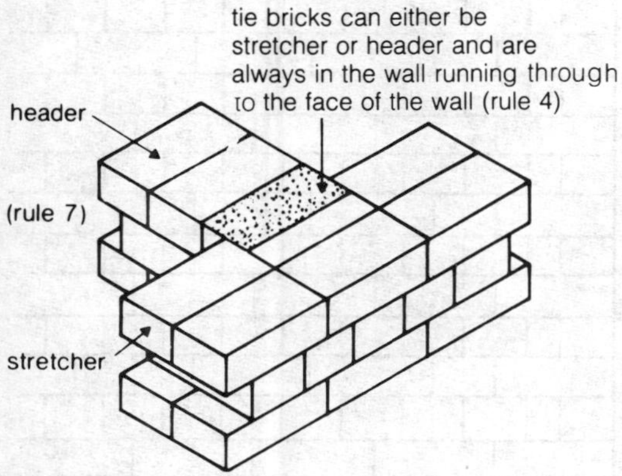

Fig. 59 Rules of bonding (rules 4 and 7)

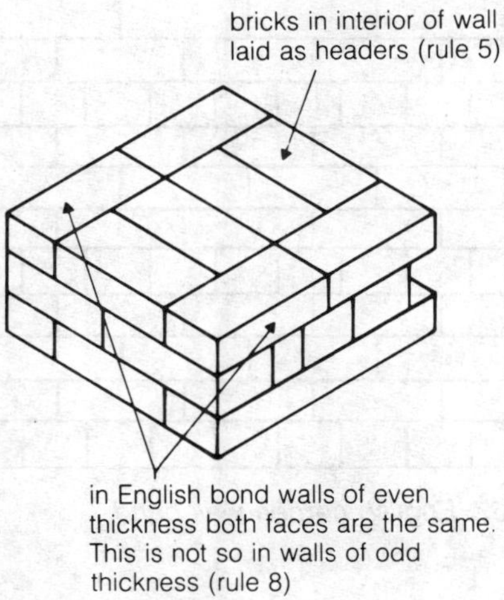

Fig. 60 Rules of bonding (rules 5 and 8)

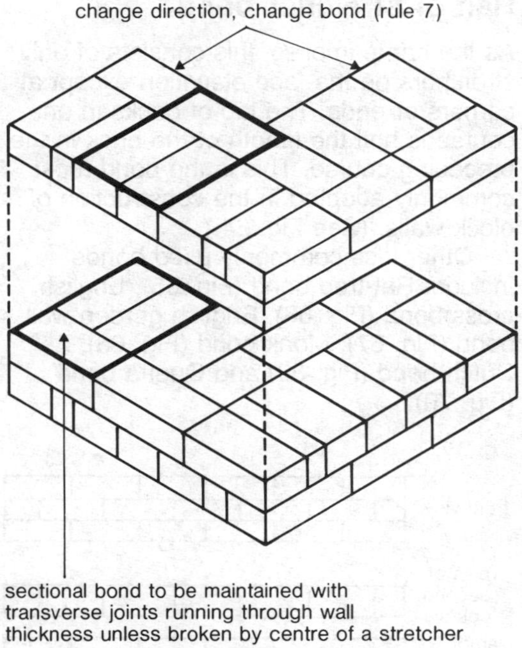

Fig. 61 Rules of bonding (rule 7)

3 No cut bricks less than a closer must appear on the face. (See Fig. 58.)
4 Tie-bricks at wall junctions which can either be headers or stretchers are always in the wall that runs through to the face of the wall. (See Fig. 59.)
5 As far as possible bricks in the interior of the wall should be laid as headers. (See Fig. 60.)
6 Unless stopped by the centre of a stretcher, all transverse joints must continue unbroken across the thickness of the wall. (See Fig. 61.)
7 When the wall changes direction, the bond must be changed in the same course, e.g. stretchers along the front and then headers on the side in English bond. The exception to this rule is in breaks of 112·5 mm or less, where two walls of differing thicknesses tie into each other. (See Fig. 61.)
8 In English bond with walls of even thickness both faces are the same. This is not so in walls of odd thicknesses. (See Figs. 60 and 61.)
9 A 'broken bond' results when the bond pattern cannot be maintained in a given length of walling. Broken bonds may be avoided by reversing the end bricks as in rule 1, when it is termed a 'reversed bond'.

Types of bond

There are many types of bond in use by the bricklayer but for our purpose we shall be concerned only with the most important. These are as follows.

English bond

This consists of alternate courses of headers and stretchers. The bond has no straight joints internally or externally and it is considered one of the strongest bonds. It is suitable for the construction of load-bearing walls and for places where strength is of utmost importance. (See Fig. 62.)

Flemish bond

This consists of alternate headers and stretchers in the same course. Flemish bond is said to give a more attractive face appearance than English bond as it appears less monotonous. It affords a saving in facing bricks because of the headers.

English bond requires approximately 89 facing bricks per square metre, while Flemish bond requires only 78 facings. (See Fig. 63.)

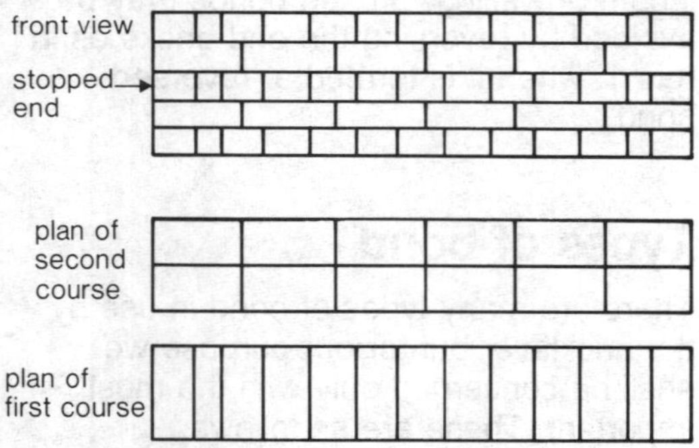

Fig. 62 English bond

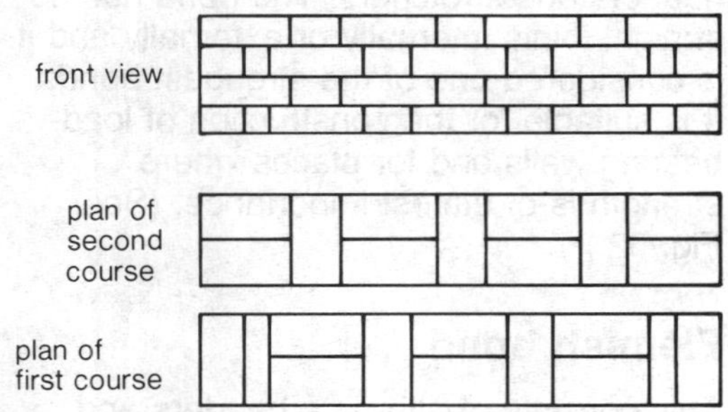

Fig. 63 Flemish bond

Half or stretcher bond

As the name implies, this consists of only stretchers on the face elevation except at corners or ends. The lap of bricks in one course is half the length of the brick in the preceding course. This is the bond most commonly adopted in the construction of block walls. (See Fig. 64.)

Other less commonly used bonds include: Rat-trap bond (Fig. 65), English cross bond (Fig. 66), English garden wall bond (Fig. 67), Monk bond (Fig. 68), Dutch bond (Fig. 69) and Quetta bond (Fig. 70).

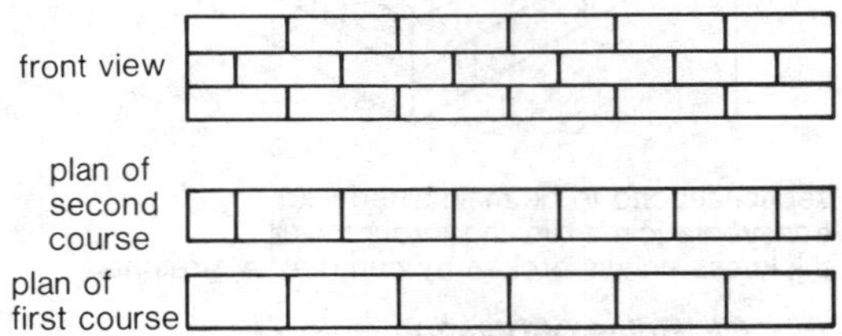

Fig. 64 Half or stretcher bond

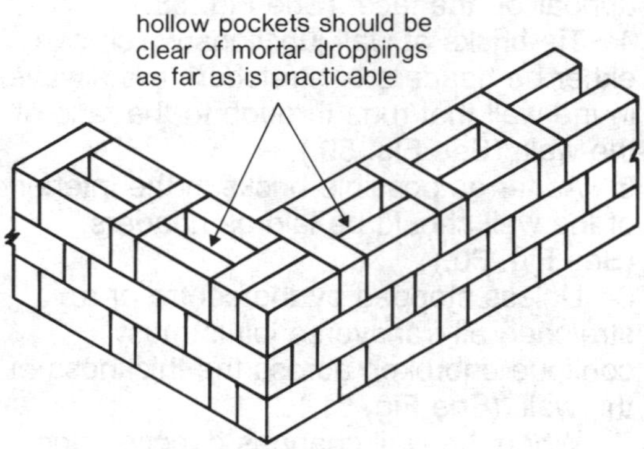

Fig. 65 Rat-trap bond

Fig. 66 English cross bond

Fig. 67 English garden wall bond

Fig. 68 Monk bond

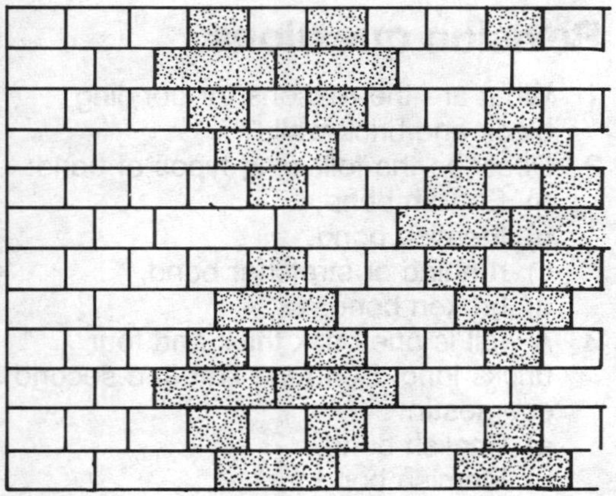

Fig. 69 Dutch bond

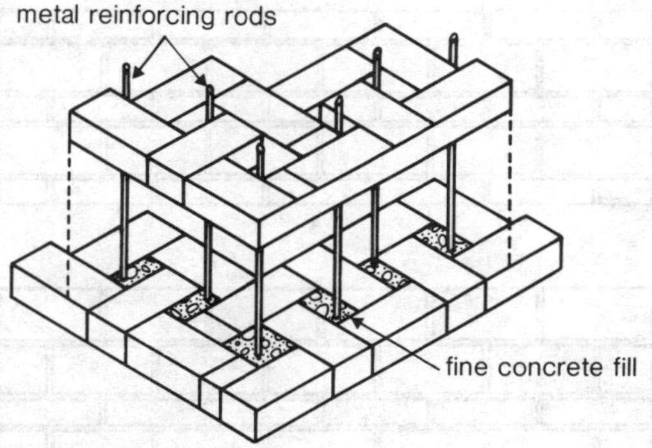

Fig. 70 Quetta bond

Stuctural and decorative panel bonds in blocks

Just as with bricks, quite interesting decorative bonds can be formed with blocks. Some of the panel wall bonds can be adopted for structural purposes provided the walls are adequately reinforced.

Stretcher bond (running or half bond)

The blocks for this bond are laid in regular courses, each lapping a half length on the blocks of the lower course. This is by far the most common block bond in use for both industrial and domestic construction. (See Fig. 71.)

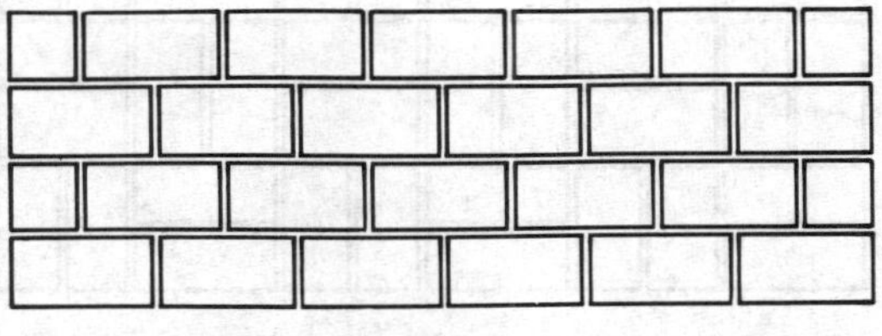

Fig. 71 Running or half bond

Stack bonds

There are three possible arrangements for these types of bonds. The blocks may either be laid as headers, as stretchers, or as soldiers on their ends. Bonds of this kind can only be used for panel infilling in framed structures. They must not be used for load-bearing walls unless adequately reinforced with steel. (See Fig. 72.)

Coursed ashlar bond

This is very attractive bond pattern obtained by using blocks of two different thicknesses. The bond is as strong as the stretcher bond. Various arrangements are possible: the larger and smaller thickness blocks can be laid in alternate courses, the larger thickness blocks can be laid at two course intervals or vice versa. In either case the smaller thickness blocks may be projected out beyond the general wall face to give an attractive profiled finish. (See Fig. 73.)

Basket weave bond

The bond, like stack bond, is not suitable for load-bearing panel walls unless adequately reinforced. Split blocks are often used for basket weave bonding. (See Fig. 74.)

Patterned ashlar bond

This is a very attractive bond and is

reasonably strong because of the tie headers. For added visual effect, the headers are projected out from the wall face. (See Fig. 75.)

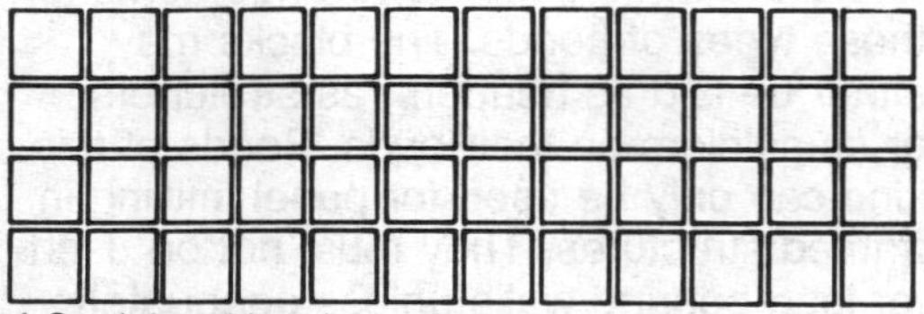

A Stack bond headers

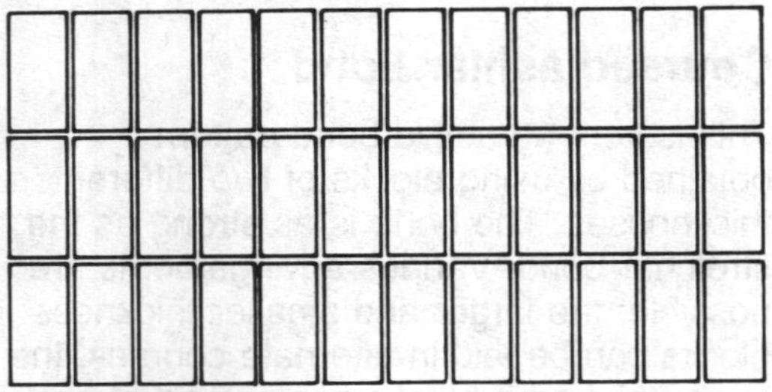

B Stack bond soldiers

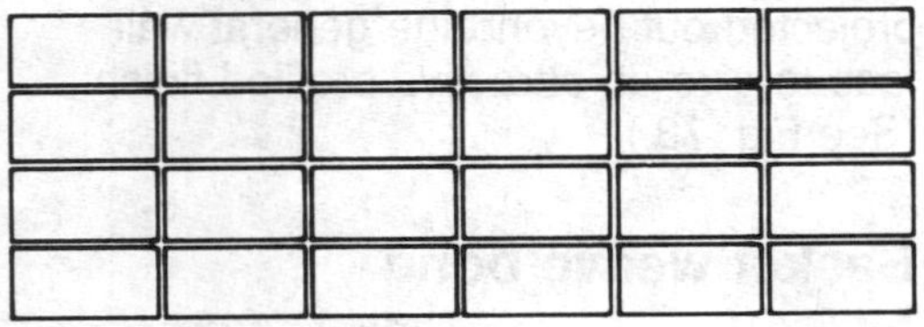

C Stack bond stretchers

Fig. 72 Stack bonds

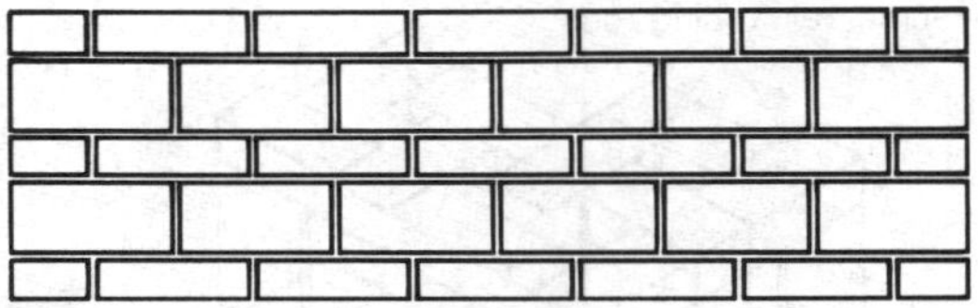

A

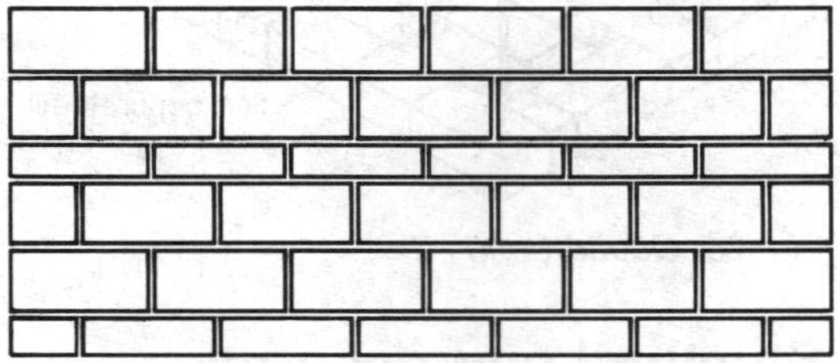

B

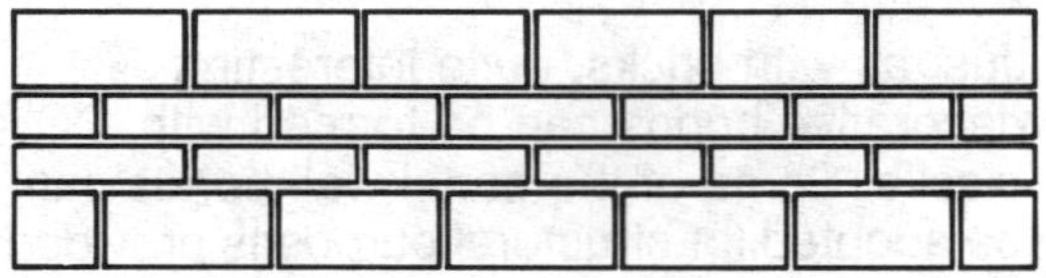

C

Fig. 73 Coursed ashlar bonds

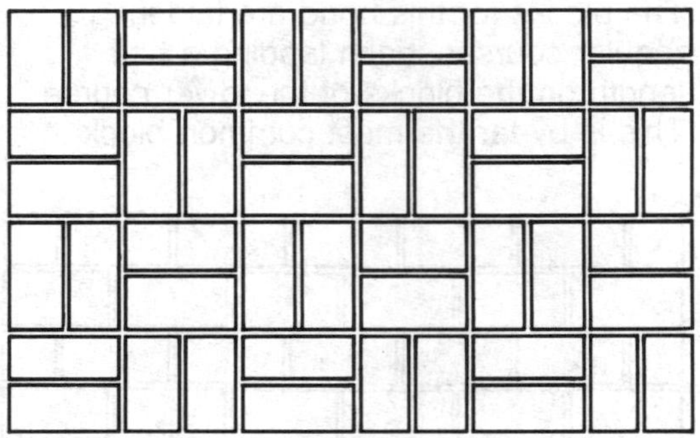

Fig. 74 Basket weave bond

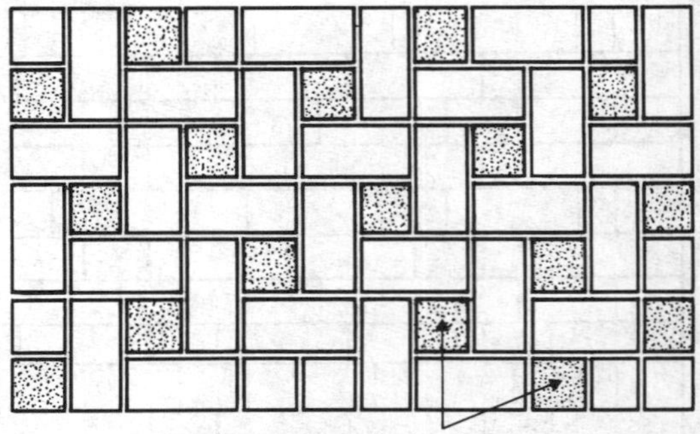

Fig. 75 Patterned ashlar bond

Revision questions

1 What are the reasons for bonding block and brick walls?

2 Describe the following types of bond:
a) English bond,
b) Flemish bond,
c) running or stretcher bond,
d) broken bond.

3 A wall is one brick thick and four bricks long. Bond the first and second courses in:
a) English bond,
b) Flemish bond.

4 With the aid of neat sketches, show the following terms in bonding:
a) toothing,
b) racking back,
c) return angle.

7. Trench excavations

After setting out the building and fixing the profile as shown in Fig. 76 the next operation will be to excavate for the foundations of the building. Excavations take various forms depending upon the type of foundation to be laid. Strip foundations require the excavation of strip trenches. Pad or isolated foundations require holes dug where the foundations are to come only, unless connected by ground beams. Rafts and basements need the excavation of the whole area of the building; this is referred to as bulk excavation.

Excavation works may be done by hand or machine. For ordinary house building hand excavation is most commonly used, while larger excavations are done with the aid of mechanical excavators.

When trenches are being excavated to a depth likely to cause the caving in of the sides they must either be given some form of temporary support or the sides must be sloped to provide self support. The support given to the sides of the trench depends upon the depth of the trench and the soil condition. Weak soils will require more elaborate temporary supports. Most temporary supports take the form of timbering the sides. As timbering is only meant to support the sides of the trench until all foundation work is complete, over-timbering should always be avoided so that progress of the work is not hampered in any way.

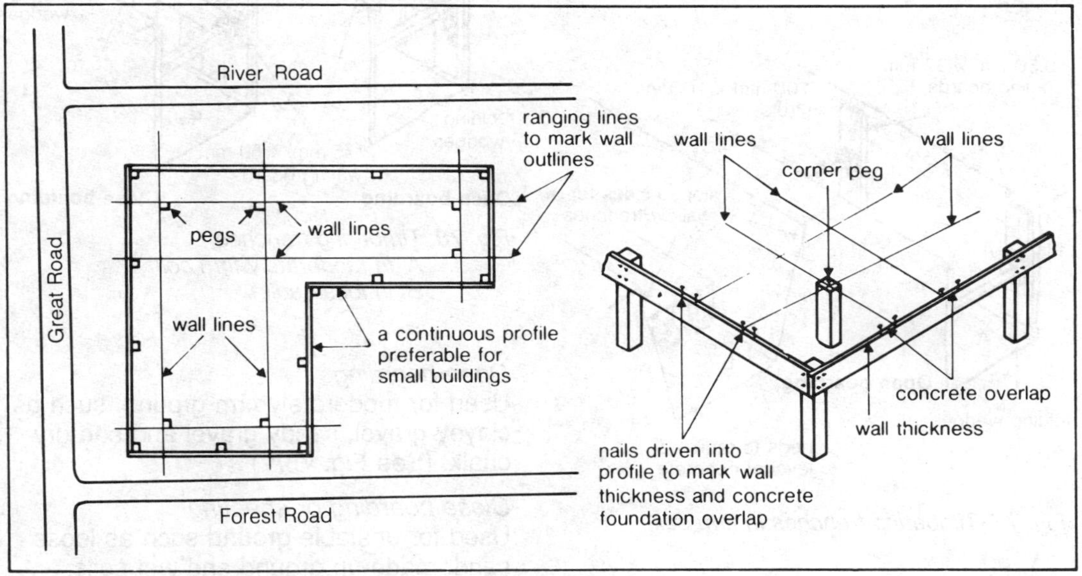

Fig. 76 Erection of profile

The following terms are used in timbering trenches.

Poling board (poling)
Vertical members usually 175 mm by 31 mm by 1 m with 1·22 m and 1·5 m lengths. (See Fig. 77.)

Waling board (waling)
Horizontal members placed against the poling boards with cross-section measurement of 150 mm by 75 mm, the length being the normal length of planks.

Struts
Horizontal supports placed across the trench to hold the other members in position. They are spaced at between 1·5 m to 1·8 m along the length of the

trench. Struts are usually 100 mm by 100 mm in cross-section. (See Fig. 77.)

Wedges
Sometimes pairs of folding wedges are used between the waling boards and the struts to take up any slackness. (See Fig. 77.)

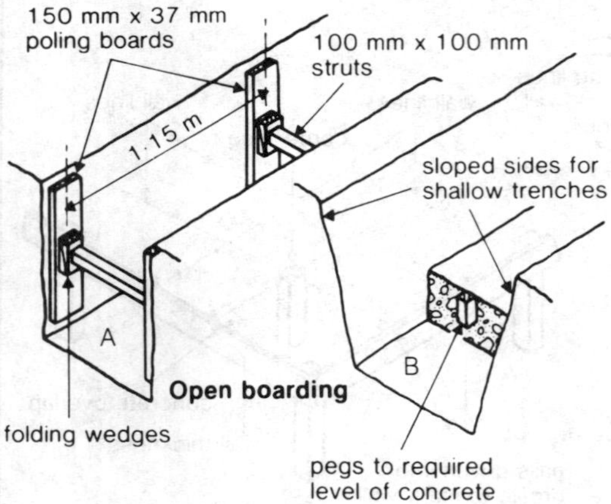

Fig. 77 Timbering trenches in firm soil

Classification of timbering

Timbering may be classified into three types as follows.

Poles and struts
Used for firm soils such as firm gravel, stiff clay and hard chalk. The poling boards are spaced 1·829 m apart. (See Fig. 77.)

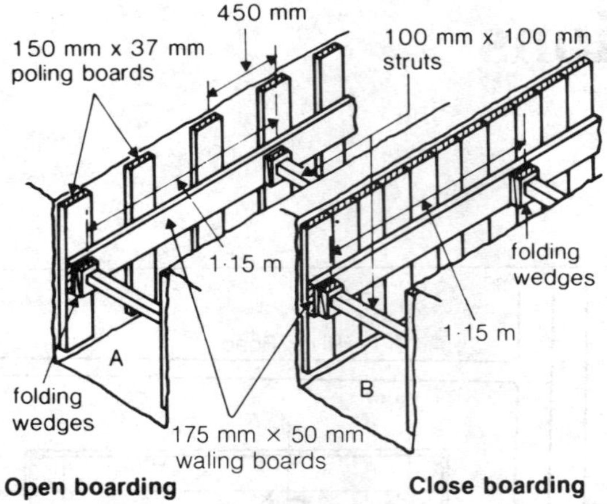

Fig. 78 Timbering trenches
A in moderately firm soil
B in loose soil

Open boarding
Used for moderately firm ground, such as clayey gravel, sandy gravel and soft dry chalk. (See Fig. 78A.)

Close boarding or sheeting
Used for unstable ground such as loose sand, made-up ground and wet soils. (See Fig. 78B.)

For safety reasons the Construction Regulations state that:
1 heavy loads or machinery likely to cause the collapse of the sides must not be moved close to an excavation,
2 open excavations must be fenced off to prevent persons falling into them (Fig. 79),
3 approaches to excavations must have warning lights prominently displayed at night,

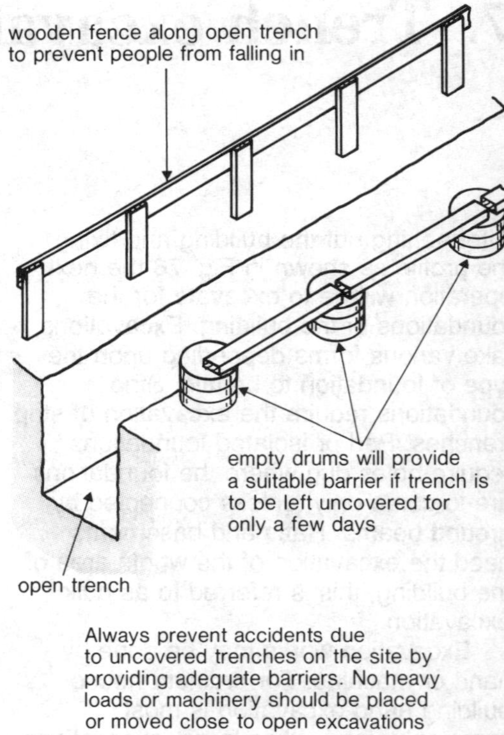

Fig. 79 Barriers to open trenches

4 no vibration likely to cause collapse of the sides must be caused close to the excavation.

Foundations

To lay concrete foundations the following procedures are followed.

Bottoming

This is achieved by cutting high portions of the bed of the excavation, filling low places and compacting the loose earth. This operation is necessary in order to ensure that the concrete rests on a flat, solid bed, thus minimising the possibility of differential settlement.

Pegging

To maintain an even thickness throughout the foundation, pegs are driven into the bed of the trench, leaving them projecting to a depth equal to the desired thickness of the concrete. During the laying, the concrete is tamped down to the level of the pegs, as shown in Fig. 77B.

Wetting

During hot, dry weather, foundation trenches must be thoroughly wetted before the mixed concrete is poured in. If this is not done, the water in the concrete will be quickly absorbed by the dry soils of the trench with the result that the concrete will develop false setting, producing a weak foundation.

Pouring the concrete

Concrete must not be poured from a height that will allow the segregation of the constituent aggregates . If the trench is deep and laying is being done manually, the concrete should be lowered in buckets and not poured directly from the wheelbarrow at ground level. After pouring, the concrete is spread with a spade and roughly tamped down to the required level. The foundation is then allowed to cure and gain strength before work is started on the walls that rest on it.

Revision questions

1 Define the following building terms:
   a) building line,
   b) datum peg,
   c) corner profile.
2 Give four safety precautions necessary when excavating a trench.
3 Produce a neat sketch of a close boarded timbering and describe the type of soil for which it may be used.
4 Describe how a newly excavated trench is prepared to receive a concrete foundation.
5 What are the precautions necessary to prevent segregation of concrete when laying a deep concrete foundation?

8. Foundations

Foundations are the lowest parts of the walls of a building. They may be formed of stone, concrete or steel, although the most common material in modern construction is concrete. The depth to which the foundations will be laid and their sizes are given by the architect and the engineer after a thorough examination of the lower layers of the soil. The pressure imposed by the foundations of a large building is considerable and where there are weak portions of the substrata, the foundations may fail during the process of eventual settlement. In areas where minerals are mined and in areas prone to excessive soil erosion, extra measures should be taken against sliding foundations.

The purpose of a foundation is to support the building, transmitting the weight of the building, both live and dead weights, to the subsoil without failing as long as the building lasts. A good foundation should not:

1 break under a load,
2 be destroyed by the roots of trees,
3 be affected by the presence of water.

The Building Regulations 1985, Part A of Schedule 1 states that foundations shall:

1 safely transfer all dead, imposed and wind loads to the ground without settlement or other movement which would impair the stability of, or cause damage to, the building itself or any adjoining building or works,
2 be taken down below frost damage or soil movement level,
3 be resistant to attacks by sulphate, or other deleterious matters found in the soil. (See Fig. 80.)

Foundations must be of sufficient thickness and be taken down to such a depth that they will be unaffected by ground movements and soil erosion

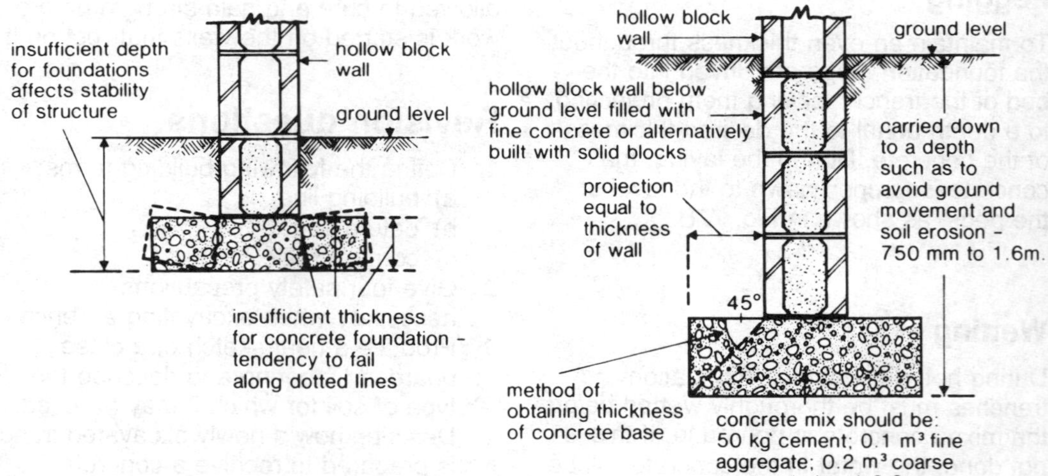

Fig. 80 The foundation

Table 4 gives the recommended minimum width of strip foundations.

Types of soil

The designer of a foundation must first of all determine the nature of the geological formation underlying the site. The method by which this is done is beyond the scope of this book. It will suffice to know the types of soil we are most frequently called upon to deal with. Soils vary

Type of subsoil		Condition of subsoil	Field test applicable	Total load of load-bearing walling not more than [kN/linear metre] 20	30	40	50	60	70
				Minimum width of strip foundation [mm]					
I	Rock	No inferior to sandstone, limestone or firm chalk	Requires at least a pneumatic or other mechanically operated pick for excavation	In each case equal to the width of wall					
II	Gravel Sand	Compact Compact	Requires pick for excavation. Wooden peg 50 mm square in cross-section hard to drive beyond 150 mm	250	300	400	500	600	650
III	Clay Sandy clay	Stiff Stiff	Cannot be moulded with the fingers and requires a pick or pneumatic or other mechanically operated spade for its removal	250	300	400	500	600	650
IV	Clay Sandy clay	Firm Firm	Can be moulded by substantial pressure with the fingers and can be excavated with graft or spade	300	350	450	600	750	850
V	Sand Silty sand Clayey sand	Loose Loose Loose	Can be excavated with a spade. Wooden peg 50 mm square in cross-section can be easily driven	400	600				
VI	Silt Clay Sandy clay Silty clay	Soft Soft Soft Soft	Fairly easily moulded in the fingers and readily excavated	450	650				
VII	Silt Clay Sandy clay Silty clay	Very soft Very soft Very soft Very soft	Natural sample in winter conditions exudes between fingers when squeezed in fist	600	850				

Note:
In relation to types V, VI and VII foundations do not fall within the provisions of this section if the total load exceeds 30 kN/m.

Table 4 Recommended minimum width of strip foundations
Source: Building Regulations 1985. Her Majesty's Stationery Office 1985

throughout the world, and may range in particle size from large stones to colloidal clay. The type of soil is, therefore, a significant factor in the determination of the type and size of the foundation.

Rock

This makes an excellent foundation. Rock is hard and durable but difficult and costly to excavate.

Coarsely grained non-cohesive soils

Under this heading, we have gravels and sand. Before building on sand, further treatment such as confining and compacting is necessary to make it stable.

Finely grained cohesive soils

These types include clay and silt which are soils with very fine particles which become sticky when wet and hold together when dry. Such soils are capable of absorbing moisture by capillary attraction, so that clay under a building can still remain damp even though there is no direct contact with water. Foundations on clay should be placed at least one metre deep.

Peat

Peat is a poor soil that can hardly bear any heavy loads. It is highly compressible and acidic.

Loam

This is often referred to as vegetable soil because of its high content of organic matter. It can carry little weight and is usually removed and used for agriculture.

Made-up ground

The strength of a made-up ground will depend on the quality of the fill material and amount of consolidation. The more usual made-up grounds are land reclaimed from marshes, sea and refuse tips. Made-up grounds are common in cities and big towns due to the scarcity of building land. Careful thought needs to be given to the selection of foundations for this type of ground. A damp-proof layer must be inserted in the floor of a building in marshy ground.

Types of foundation

There are many types of foundation, each one being selected according to the type of condition and the load expected to be carried. For our purpose, it will suffice to limit our discussion to the common ones with which we are bound to deal. These are:

a) strip foundations,
b) pad foundations,
c) raft foundations,
d) pile foundations.

Strip foundations

Strip foundations, as the name implies, are continuous strips of mass concrete or reinforced concrete or stones laid at a pre-determined depth below ground level and along the positions of the load-bearing walls only. These types of foundation are suitable for boundary walls, retaining walls and domestic buildings of not more than three storeys. They are not suitable on made-up ground, very soft clay, silt or peat. A strip foundation may be stepped in sloping ground to save excessive digging at the higher level. The construction details are given in Fig. 81A. Figure 81B shows sections through strip foundations in concrete and stone.

Footings

Footings, which are the enlargements of the bottom courses of the wall, may be used in conjunction with strip foundations with the aim of spreading the imposed loads to a greater area of the soil. They are rarely used nowadays, probably because of their cost and the fact that stronger and better concrete is now produced.

Different localities may have different regulations regarding the sizes of foundations. The rules that follow are based upon generally accepted practice and comply with existing regulations.

1 Where no transverse reinforcement is provided, the thickness of strip foundations should not be less than 150 mm.

2 The width of the concrete foundation should be equal to twice the width of the wall plus 200 mm.

3 If footing courses are used the depth may be obtained graphically, and where

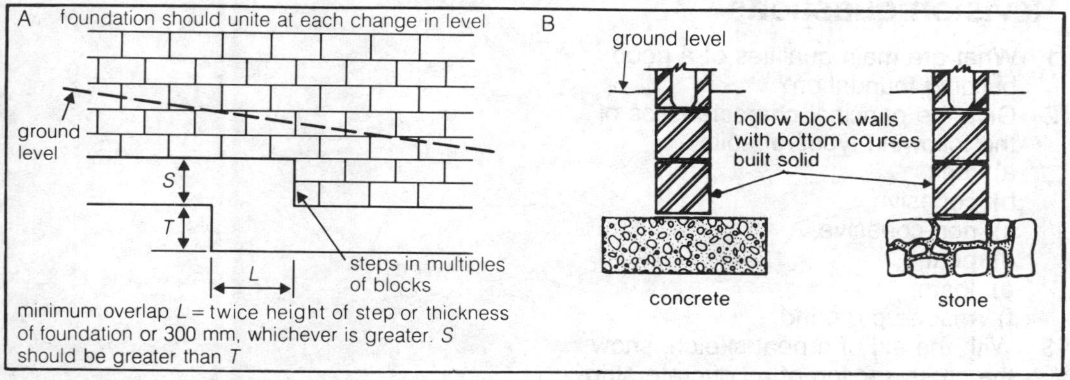

Fig. 81 Strip foundations

no footings are used the depth should be equal to 1⅓ times its projection from the face of the wall.

Pad foundations

These may be circular, rectangular or square in section. The most common types are square. They may be of mass or reinforced concrete, but reinforced concrete pads are reserved for the larger types of structures. They are generally used to support isolated loads such as those in columns, piers and heavy machinery in factories. Figure 82 shows a simple pad foundation.

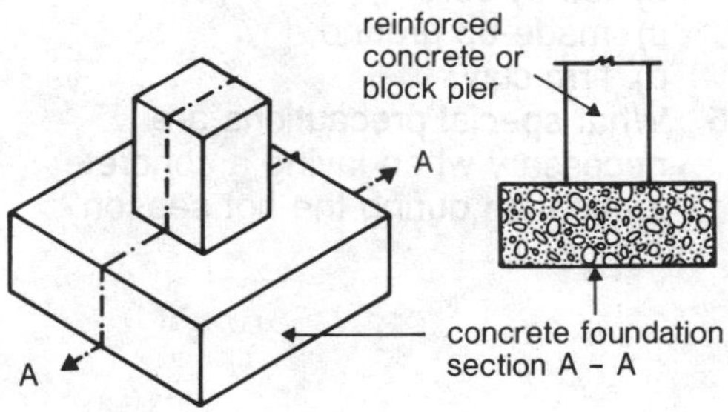

Fig. 82 Pad foundations

Raft foundations

The name for this foundation is derived from the raft which is used to cross a river. The foundation consists of a continuous reinforced concrete slab under the whole building. The foundation uses the principles of a floating raft and the weight of the building is evenly distributed over a large area so that no particular area is made to receive heavier loads. For ordinary housing purposes, a raft of 150 mm to 225 mm thick reinforced concrete is sufficient; this thickness can be increased to from 225 mm to 375 mm for slightly larger buildings. Raft foundations may be used together with ground beams for greater load-bearing capabilities. (See Fig. 83.)

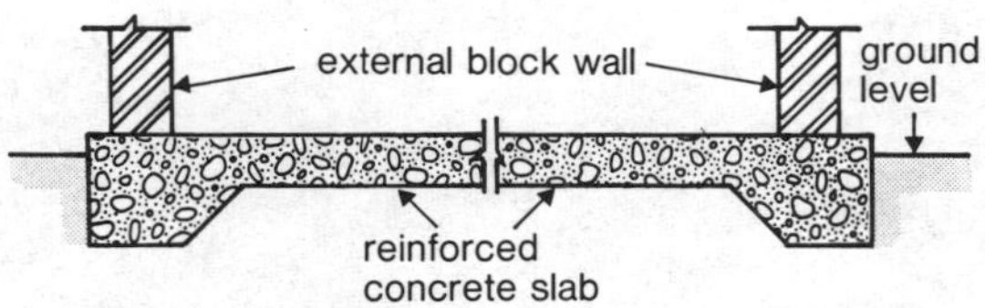

Fig. 83 Raft foundations

Pile foundations

The construction of pile foundations is beyond the scope of this book. We shall therefore limit the discussion to short bored piles which are used for transmitting loads from a building to a bearing stratum where such a stratum underlies a depth of weak upper soil. Short bored piles are circular in section, from 250 mm to 350 mm in diameter, and are used where the stable soil lies at a depth not greater than three metres. It is more economical under this type of condition as less digging and timbering is required than would be necessary with other kinds of foundation.

Mechanically or hand operated augers of the correct diameter are used to bore the holes into which the piles are cast. Reinforcing rods are usually introduced to the top part of the piles and linked to those of the ground beams which come on top of the piles. Figure 84 shows the plan and section of short-pile building foundations.

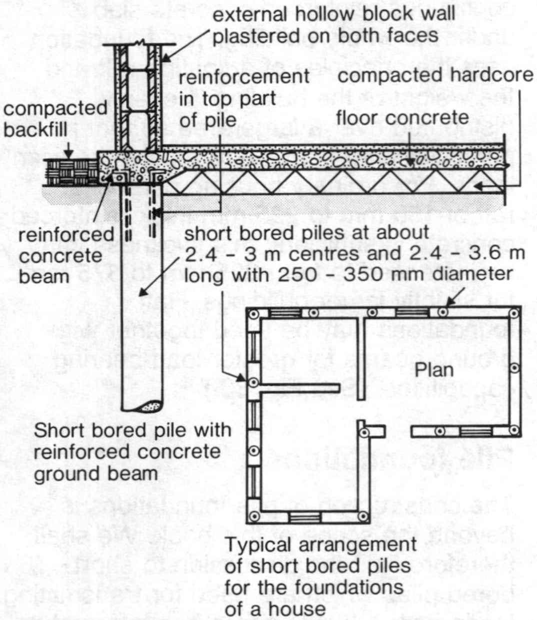

Fig. 84 Short bored pile foundations with plan for a typical small house

Revision questions

1 What are main qualities of a good building foundation?

2 Give the principal characteristics of the following types of soils:
a) rock,
b) cohesive,
c) non-cohesive,
d) peat,
e) loam,
f) made-up ground.

3 With the aid of a neat sketch, show the cross-section of a concrete strip foundation.

4 Describe the most suitable types of foundation for the following:
a) sandy soil,
b) loamy soil,
c) made-up ground,
d) firm clay.

5 What special precautions are necessary when laying a concrete foundation during the hot season?

9. Wall construction

Materials layout

The correct layout of materials on the building site (Fig. 85) is one of the first steps towards achieving economical block walling. Advantages include:

1 speed in blocklaying as the mortar, blocks and tools are conveniently placed within easy reach,

2 a safer site is ensured,

3 higher standard of workmanship is achieved,

4 wastage through breakages is minimised.

Blocks, when not manufactured on site, must be bought from reputable manufacturers. Whatever the source of supply, they must be well cured before being used on the job. The use of 'young' blocks often results in cracking of the wall due to drying shrinkage. It can also cause the 'erosion' of external angles of walls, leaving ugly finishes when the walls are not plastered. The time taken for blocks to cure properly and dry ranges from 20 days to one month, depending upon the method adopted in curing.

No effort should be spared to protect the arrises of the blocks during the loading, transporting and off-loading. The amount of care exercised will be dependent upon the shape and size of the blocks. Packaging or palletting should

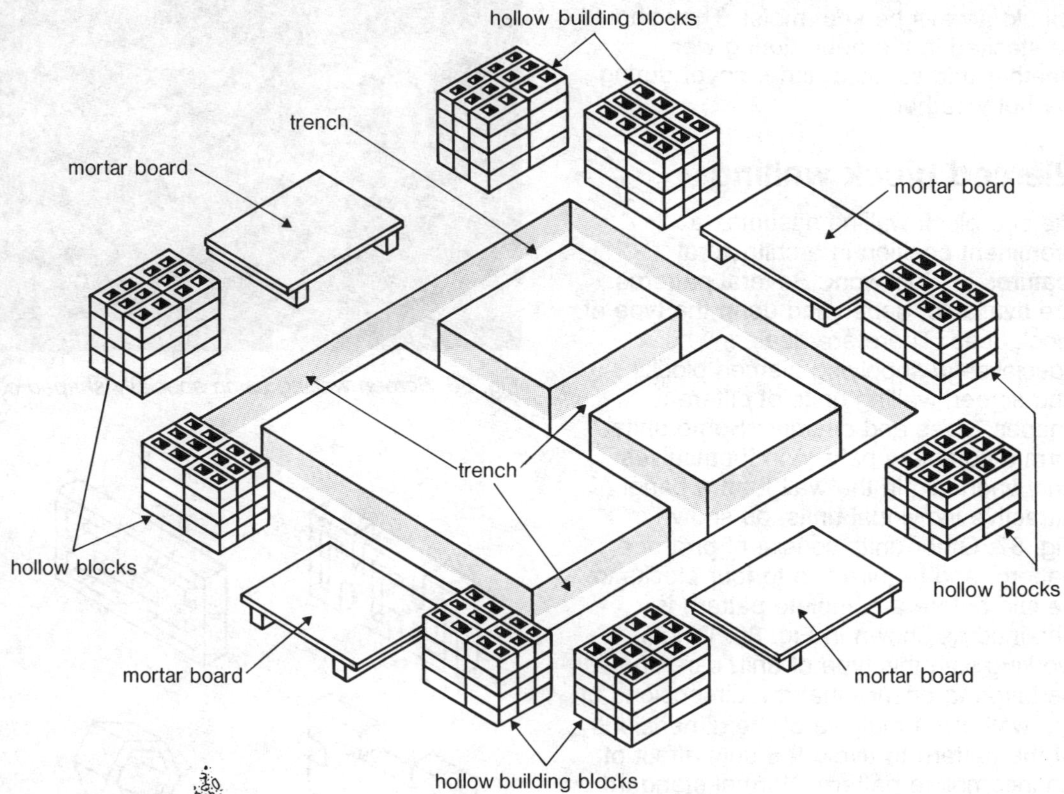

A well laid out site speeds up operations. It is safe and economical. Material layout and general planning should not be considered a waste of time

Fig. 85 Materials layout

be the system to adopt when blocks have to be transported a long way.

When on site, blocks should be stacked close to where they will be used. To prevent unnecessary damage they should be stacked on their bedding faces on a hard level surface. To ensure adequate strength of the wall, blocks should always be kept moist. They should be stacked in the open during wet weather and stacked under cover during dry hot weather.

Pierced block walling

Pierced block walling assumes a prominent position in architectural features in blockwork. Several patterns are available, depending upon the type of block used. There are dealers who specialise in supplying pierced blocks and screen walling units of different shapes, sizes and designs. Some units form a complete pattern in themselves and when laid in the wall form a panel of attractive individual units, as shown in Fig. 87. Other units consist of part of a pattern, and require two to four blocks to be laid before a complete pattern is obtained as shown in Fig. 88. When working with this type of unit, care should be taken to ensure that the dimensions of the wall are a multiple of the dimensions of the pattern to avoid the ugly effect of an incomplete pattern. Normal standard blocks, solid or hollow, can be used quite successfully in screen walls. Solid blocks laid with spaces in between them give a pleasant honey-combed effect, while hollow blocks laid on their sides or ends,

Fig. 86 Screen walling using specially shaped concrete blocks. A variety of designs is possible

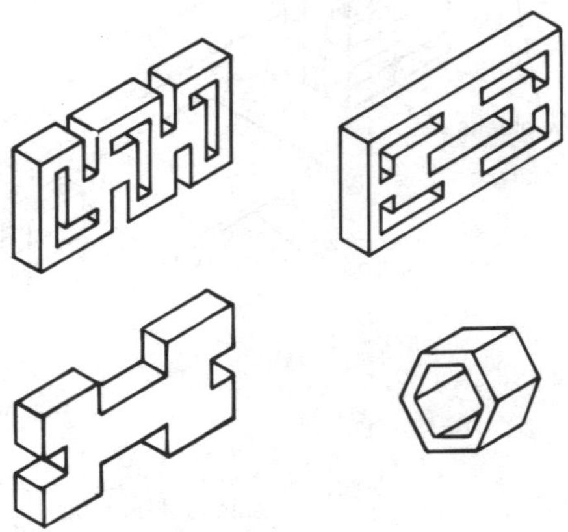

Fig. 87 Pierced and screen walling: units that are a complete pattern in themselves

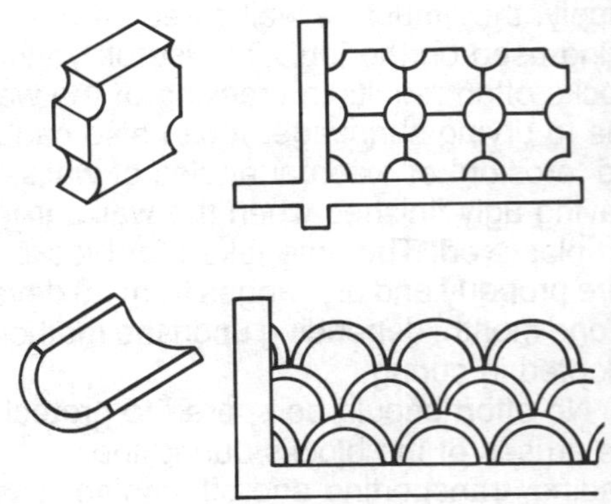

Fig. 88 Pierced and screen walling: units requiring two to four blocks to be laid before a pattern is formed

also provide an interesting and attractive screen wall. These are shown in Fig. 89. Evenness and regularity should be the guiding words when screen walling or using pierced blocks to form patterns in panel walling. Irregularity and unevenness distress the eye.

The principal function of pierced blocks is to diffuse strong sunlight and yet permit free flow of air. This dual advantage makes the use of this type of block popular in tropical climates. A major disadvantage is that the many holes created in the wall provide easy resting places for dust and dirt, snakes and lizards. This can make the maintenance costs of screen walls high.

Setting out the first course of walls

After the concrete foundation has set reasonably well, usually 24 hours after casting, the first course of wall can be set out. Lines are strained from the wall marks on the profiles. Starting from the external corners of the building, transfer these lines to the foundation. This can be done by using either a plumb bob or plumb levels. Mortar is laid to the mark and the corner block bedded. This process is shown in Fig. 90. After these operations are complete for all the corners, the walls at the corners are then raised, using a gauge rod or water level to obtain a common level. Not much attention is paid to the finishing because the walls are below ground level,but it is necessary to keep the courses level

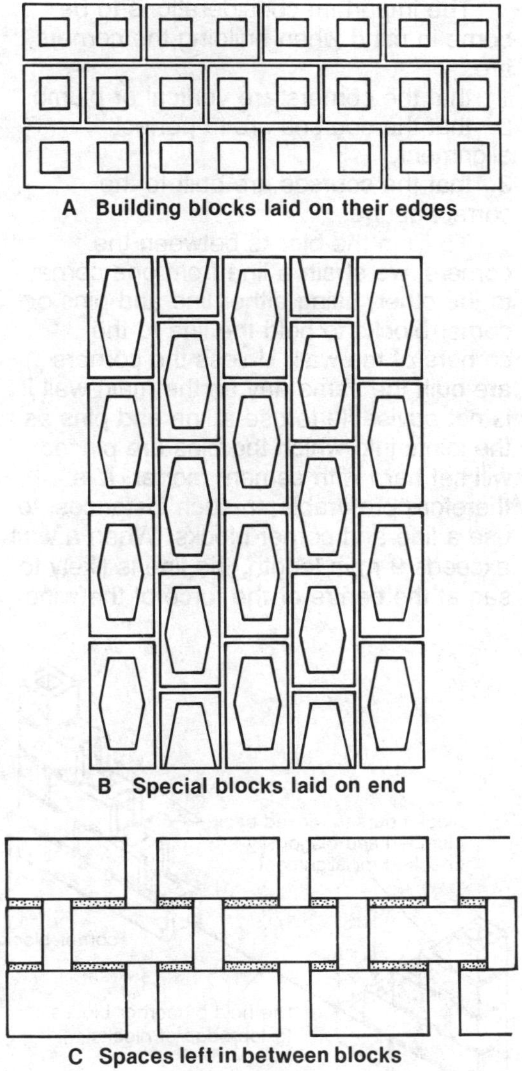

Fig. 89 Pierced and screen walling: special arrangements

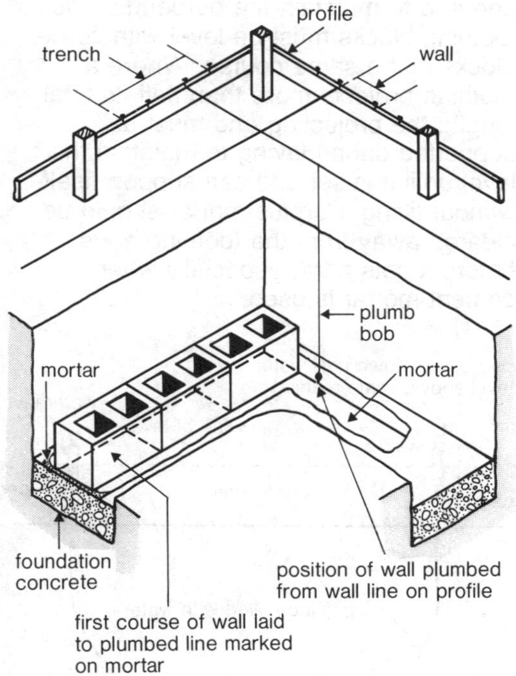

Fig. 90 Setting out the first course of a wall

throughout as this affects the neatwork (i.e. the work above ground level). An illustration of the use of a water level is shown in Fig. 91.

The raised corners are either toothed or racked back to facilitate the joining of the courses to the corner and to ensure strength and the maintenance of the original bond. When racking back corner courses, it is necessary to ensure that blocks in the off-sets in each course are in perfect alignment when checked diagonally with a straight-edge. Toothed courses should be plumbed at the face and end to maintain the perpends. The toothing blocks must be level with corner blocks in the same course. Where a toothing projects more than half its total length, the projecting end must be supported during laying to maintain the level until it is set and can support itself without tilting. Surplus mortar should be cleared away from the toothing ends before it sets hard, especially when cement mortar is used.

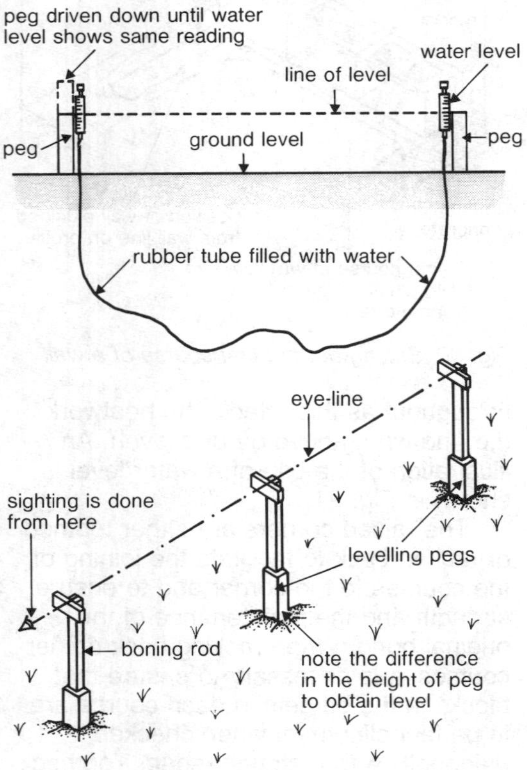

Fig. 91 Site levelling

The important considerations to be borne in mind when building the corners are:

1 that the corners are vertical or plumb,
2 that the courses are in perfect alignment,
3 that the courses are built to the correct gauge.

To fill in the blocks between the corners, we strain a line from one corner to the other, using either line and pins or corner blocks to hold the line to the corners of the wall. Unless the corners are built the same day as the main wall it is not advisable to use a line and pins as the joints into which the pins are placed will set hard with cement mortar. It is therefore preferable, in such instances, to use a line and corner blocks. When a wall exceeds 9 m in length, the line is likely to sag at the centre or the force of the wind may put it out of alignment. It is necessary, therefore, to use a tingle plate to take up such a sag. These points are illustrated in Fig. 92. When laying the blocks to the line, ensure that a trace of daylight is left between the line and the edges of the blocks. This is to prevent the line from becoming out of alignment by laying the blocks tight to it. A tingle is effective up to lengths of approximately 12 m. Beyond this length, it is advisable to divide the wall by erecting a 'lead' or pier.

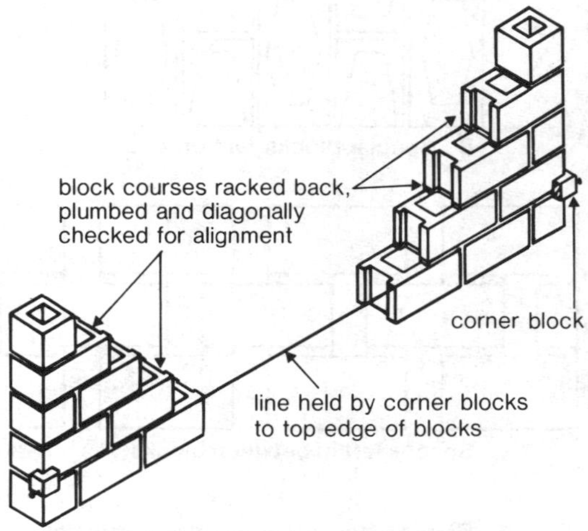

Fig. 92 Block walling

Fig. 93 Building a long block wall. Note the stacking of blocks along the wall under construction

Setting out openings

Once the wall reaches ground level thought must be given to the positions of window openings that will be formed in the wall later, and the arrangement of the blocks in the chosen bond. The drawing should be properly studied well in advance so that all odd pieces of block and broken bonds are arranged to appear directly below the positions of the openings. When building with blocks in running bond little difficulty is experienced in setting out the bond for openings. With brickwork, however, necessary precautions must be taken so that the bond at ground level is continued throughout the height of the wall. This means that the perpends must be maintained through the height of the wall.

Pre-cast window sills are built into the walls as the work proceeds. Where a wall has several window openings, it is good practice to bed the end sills first, and align all other sills to these. The bottom edge of the sills must give a true eye-line when finally bedded. If sills are to be cast in-situ, adequate provision must be made to tie them into the wall. In brick walls, bricks equal in thickness to the sill are bedded in sand at the reveals, which are later removed. Blocks need to be cut to shape to provide for the sill before they are built into the wall. Figure 94 shows pre-cast window sills in position.

Window frames may be fixed into position as the work proceeds or fixed after the walls have been completed. Whichever way is adopted depends upon several factors, for instance, the availability of the frames at the time the wall is being built may be a deciding factor. Each method has its advantages and disadvantages. In theory, it is argued that the fixing of window frames while the work proceeds enables the frames to be fixed rigidly to the walls by using fixing lugs, cramps and similar devices. A major disadvantage in this method is that the frames need to be supported in position until the walls are built. This slows down the pace of the blocklayer as the bracing

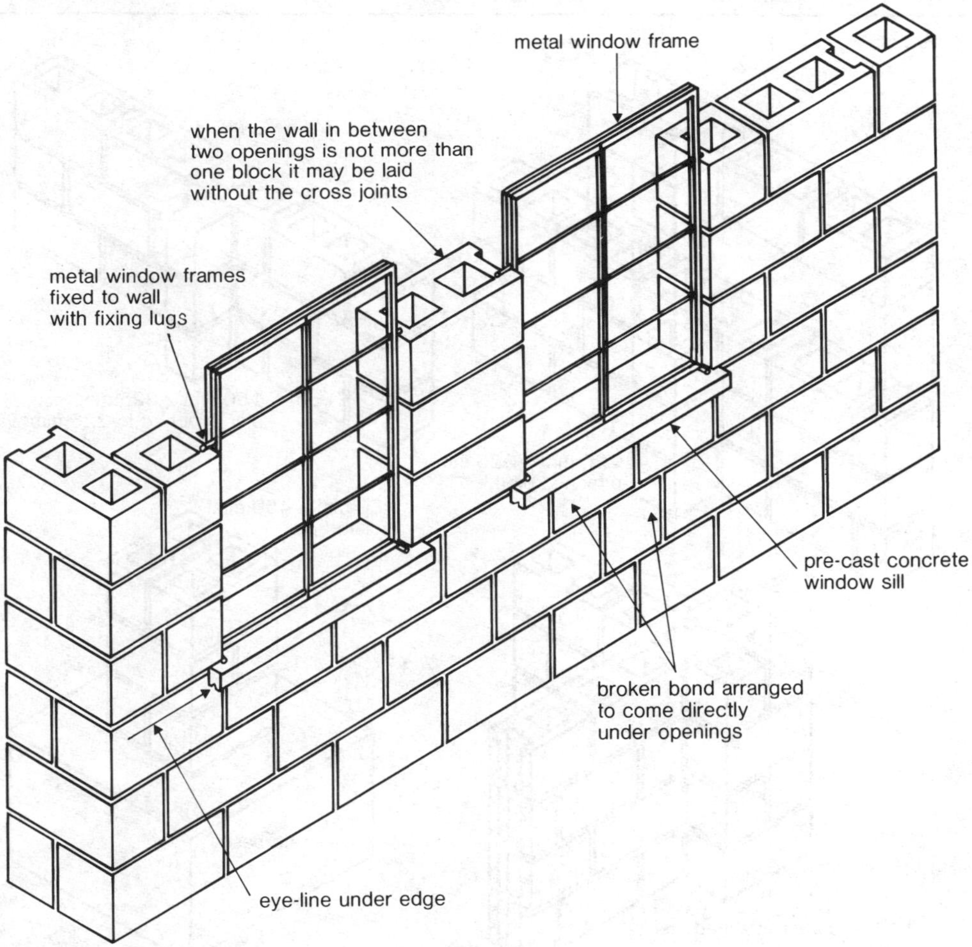

Fig. 94 Setting out openings

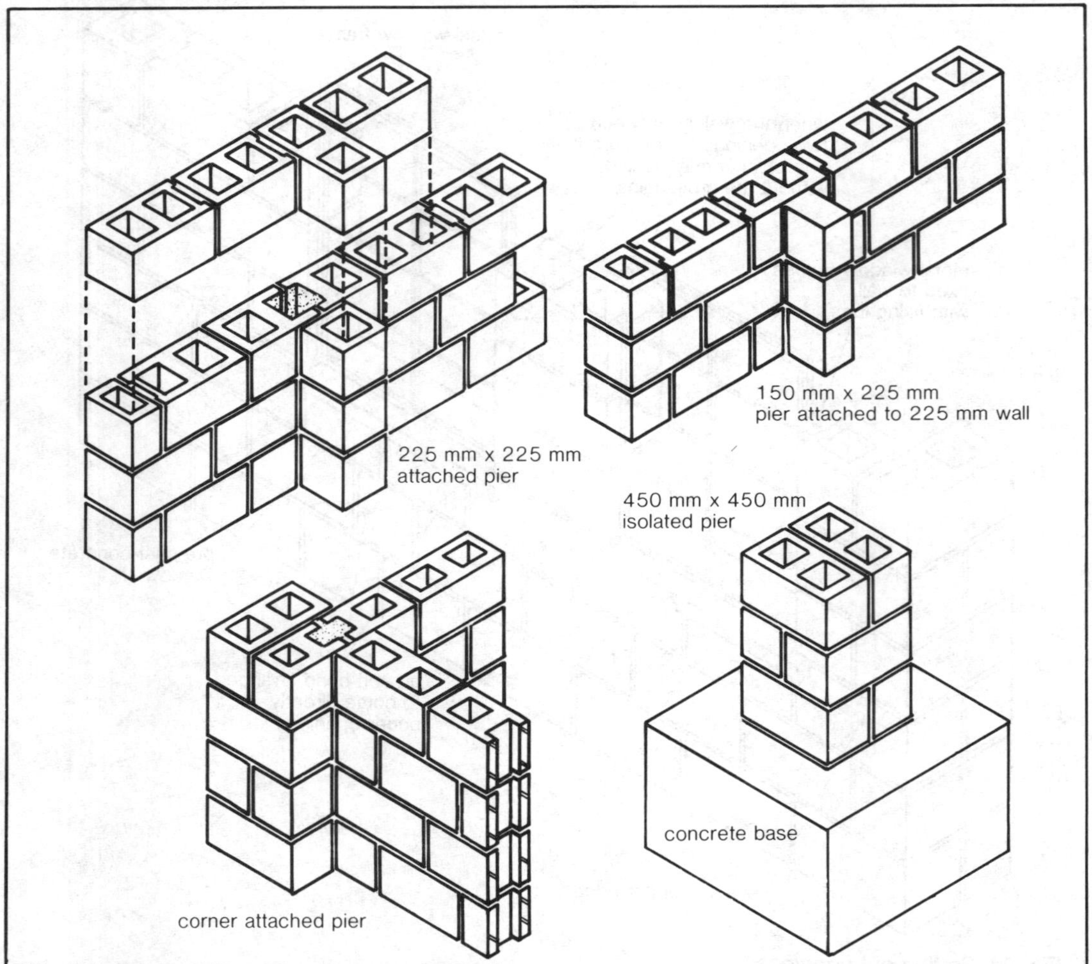

Fig. 95 Block wall construction – craft practice

timber often extends to the scaffold platform, thereby obstructing the passage of wheelbarrows and labourers.

In practice, it has been found that frames fixed after the walls have been built do not suffer any structural weakness. After all, the plaster on the reveals serves to keep all frames rigid enough. When frames are not fixed as the work proceeds, the window reveals must be kept plumb.

A 'pinch rod' is used to keep a constant check.

Figures 95 and 96 show further details of block wall construction.

Lintels and arches

These are introduced into the wall construction to span openings, usually door and window openings. Their design is based upon the usual design principles for structural parts of the building, i.e. the span of the opening, the load to be carried, etc.

Lintels

Lintels are rectangular in shape. They may be of reinforced concrete or wood. The width is normally the same as the thickness of the wall, while the depth varies according to the width of the opening. A depth or thickness of 225 mm is adequate for lintels spanning openings of up to 2·3 m wide.

The reinforcements in concrete lintels are placed between 25 mm and 37 mm from the under surface of the lintel to take up the stresses set up in the lintel when loaded. The ends of the reinforcing rods are cranked into hooks for better anchorage in the concrete. The details of this cranking are given in Fig. 97.

The number of reinforcing rods is determined according to the width of the lintel. Usually there is a 12 mm diameter rod for every 112·5 mm of width of lintel. This gives two rods for a 225 mm wide lintel.

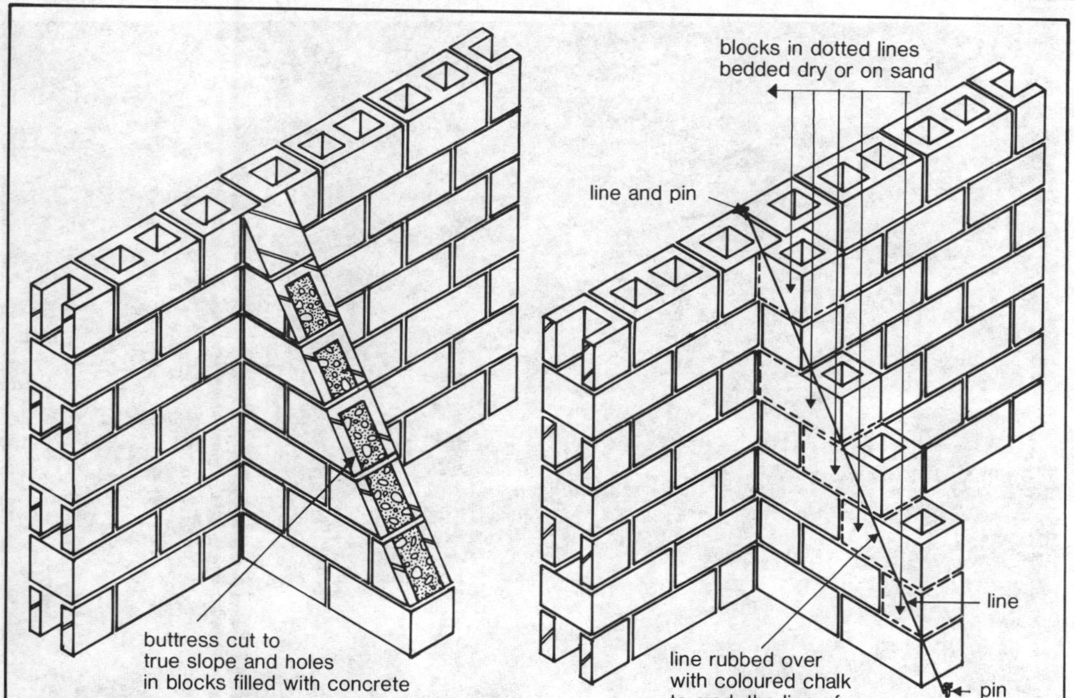

Fig. 96 Block wall construction – craft practice

pushing the upper courses over and the lintel falling on to bystanders.

A cast-in-situ lintel has the advantages of eliminating the danger involved in lifting a pre-cast lintel into position. It also bonds better with the wall as the concrete is cast wet into position. It may however be more expensive due to the formwork with all the props and braces which become necessary in supporting the wet concrete until the final strength is achieved.

To cast the lintel, prepare the mould by oiling its inner sides. This is to prevent the dried concrete sticking to the sides when striking. The formwork, if cast-in-situ,

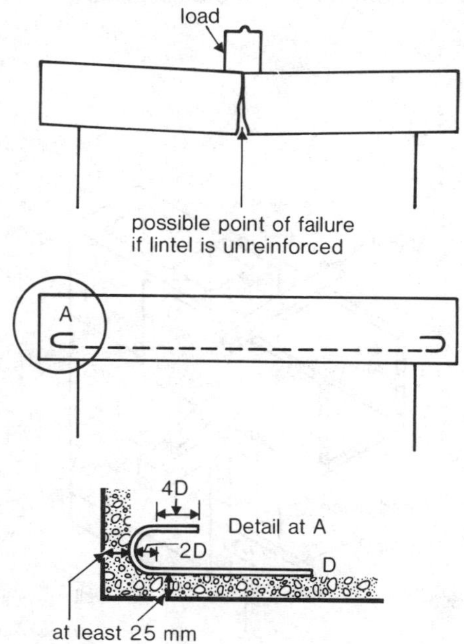

Fig. 97 Spanning openings with no steel reinforcement

Casting a reinforced concrete lintel
A concrete lintel may be cast on the ground and left until it gains sufficient strength before being built into the wall. This type is known as pre-cast. Alternatively it may be cast wet into the final position in the wall with the aid of formwork and left in that position to gain strength. This type is known as cast-in-situ.

The advantage of pre-casting a lintel is that no major formwork is required; a simple mould is all that is needed. Also, being at ground level, it is more convenient to work with and a better product is obtained. There is no danger of collapsing formwork and walling can progress uninterrupted, as there is no need to wait for the lintel to set and gain strength. A major disadvantage, however, is that the lintel is heavy and must be lifted into position. Not many firms possess lifting equipment such as hoists, cranes, pulley blocks, etc. It is very dangerous to lift a heavy concrete lintel into position. Whenever it is done, ensure that the wall on which it is to be placed has set properly to eliminate the danger of

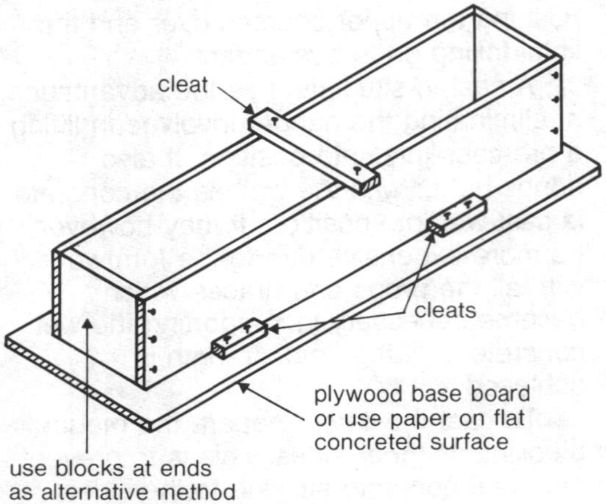

Fig. 98 Formwork for a pre-cast lintel

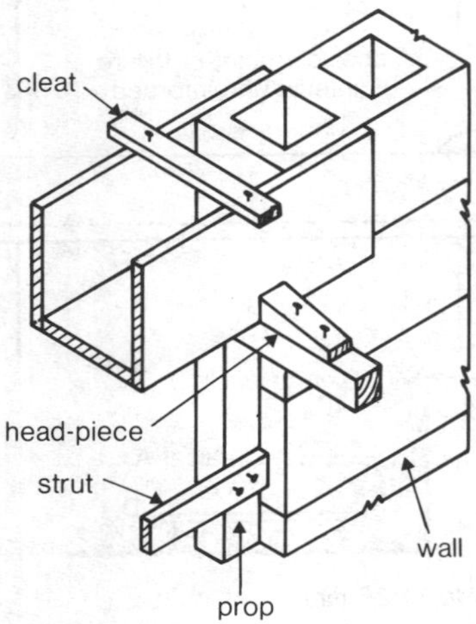

Fig. 99 Formwork for a cast-in-situ lintel

Fig. 100 Column formwork ready to be placed in position

should be designed for easy striking, fixed rigidly and adequately supported to prevent collapse or deformation. A concrete mix of 1:2:4, high quality cement, sand and coarse aggregate respectively, is prepared.

First, place a layer of concrete 25 mm thick in the mould, insert the reinforcement and then fill in the mould. Vibrate thoroughly. Pre-cast lintels must have their tops clearly marked. Sufficient time must be allowed for cast-in-situ concrete lintels to set before striking the formwork. A guide for the minimum times required before striking formwork is given in Table 5.

Formworks are also required for casting upper concrete floor slabs, beams, walls and columns. Figures 101 and 102 show column formworks with two different forms of spacing the yokes or clamps. Fresh concrete exerts a great deal of pressure inside the formwork. This pressure is greater at the lower portion of the formwork than the upper portion, if the concrete is placed in one pour. To counteract this differential pressure at the lower portion, variable yoke spacing, as shown in Fig. 102, is required. However, if the concrete is placed in batches, the pressure evens out and regular yoke spacing as shown in Fig. 101 can be used. A column of height 2·7 m or less is described as short and it can be filled with concrete in one pour from the top. Columns taller than 2·7 m, however, cannot be poured from the top because it is difficult to do so and a satisfactory consolidation of the concrete cannot be assured. Tall columns are usually poured in batches.

Lintel	**Minimum striking time at 28 °C**
Short span lintel not exceeding 1·6 m	Sides – 1 day Soffit – 5 days
Long span lintel exceeding 1·6 m	Sides – 1 day Soffit – 14 days

Table 5 Minimum striking times for cast-in-situ lintel formworks

Fig. 101 Column formwork being set in position

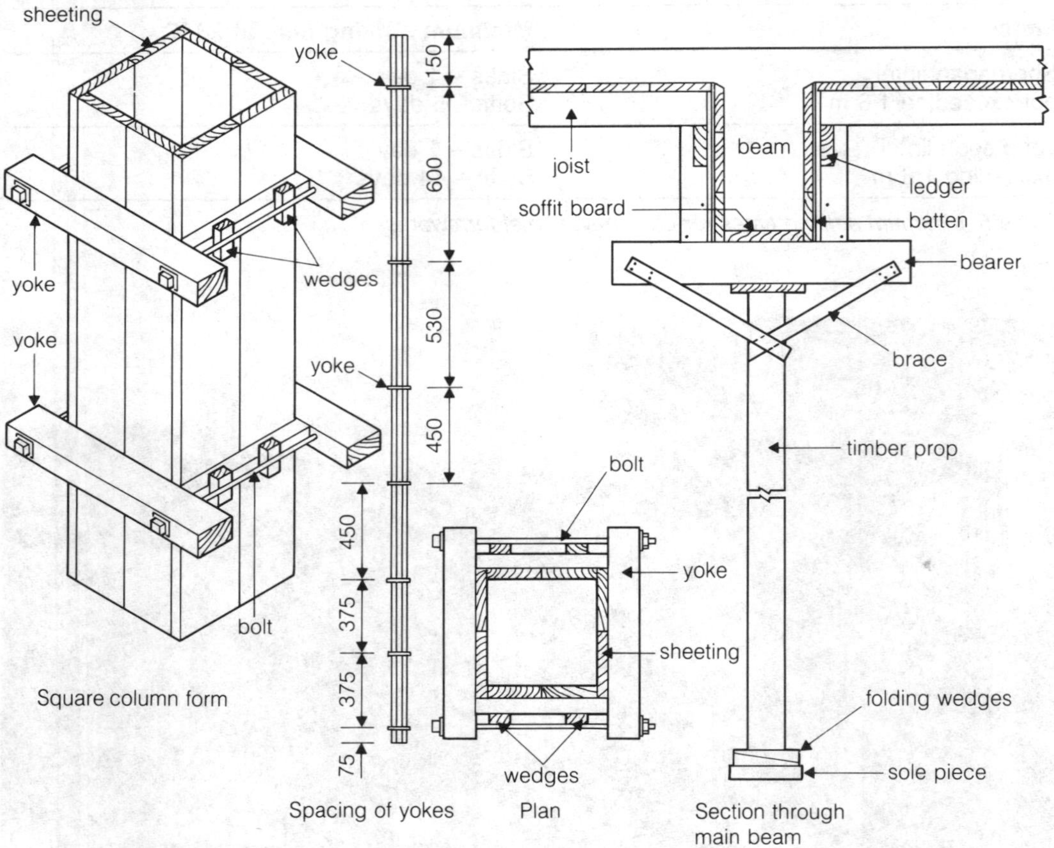

Fig. 102 Formwork details

Timber is also used for lintels. In fact timber was extensively used for spanning openings in the early history of building but the advent of concrete has reduced its popularity considerably. Timber is still used for spanning openings in house construction in many parts of the world today. It is used not only in rural areas, although this is where it is most common, but also for dwellings in urban areas too. The system of using a flat plank of thin section is not advisable, as it does not provide adequate bearing capacity for the loads imposed upon it. Often this type of lintel fails under load, causing a serious deflection of the top rail of the door frame. Wood lintels must be of adequate thickness for the imposed load. The ends of lintels must lap well on the walls supporting them. A lap of at least 150 mm should be allowed.

Arches

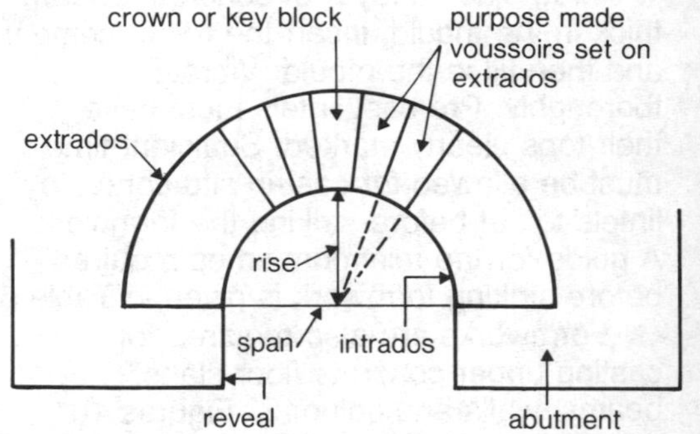

Fig. 103 Terms used in arches

In the past these consisted of separate units arranged over an opening in which the joints radiated from the springing point and the units held together by mutual pressure, thus enabling them to carry loads. This definition does not, however, stand true in the present day as modern construction techniques have permitted an arch to be conveniently cast in one whole using concrete. Arches take various geometrical forms; they may be semi-circular, semi-elliptical, segmental, or pointed. Brick arches provide very attractive finishes over openings, but can only be constructed by highly skilled craftsmen.

An arch may be constructed of specially made concrete units as shown in Fig. 105.

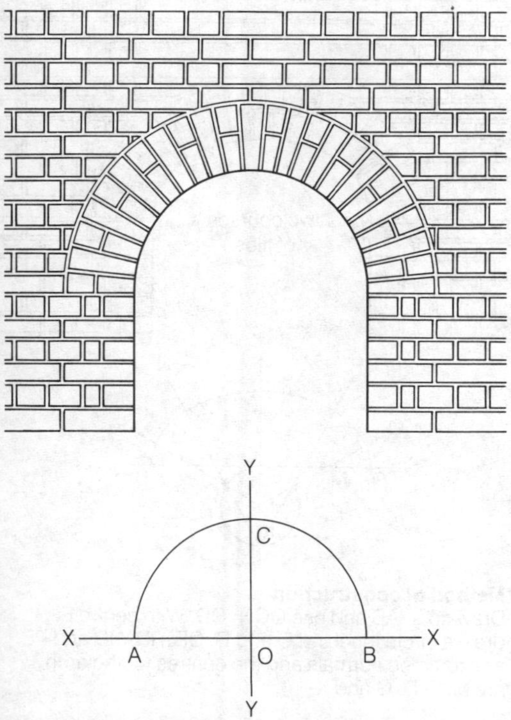

Method of construction
Draw lines X–X and Y–Y, at the intersection O, fix the point of the compasses and open it to point A or B or C, which is half the radius and equals the rise. With the point of the compasses still at O, describe an arc passing points ACB which is the intrados of the arch.

Fig. 104 Semi-circular arch

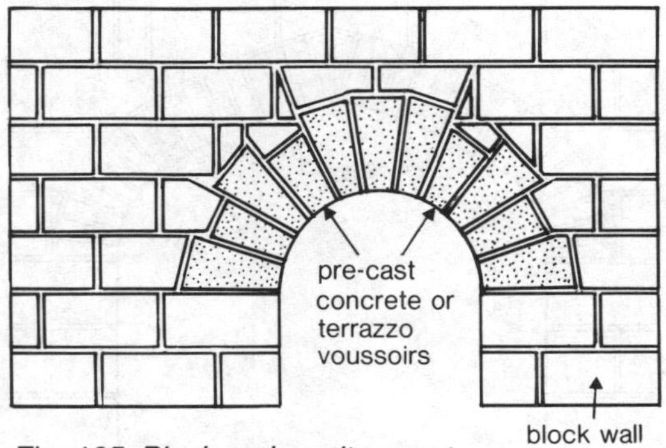

Fig. 105 Block arch: units uncut

Centre and turning pieces

These are temporary wooden supports shaped to the profile of the arch soffit to support the units during building operations. The appearance of the finished arch is highly dependent upon the type of support provided at the time of building in the units. A centre should be made of a high quality timber and be strong enough to carry the weight of the arch material and the live weight imposed at the time of construction without failing. It should be made simply with the minimum of material and labour consistent with safety. It should be made and arranged in such a way that after the arch is completed, it can be gently withdrawn without causing any vibration to the finished work. The laggings of the centre should be spaced according to the type of material used for the arch. Brick arches therefore require the laggings to be reasonably close together. The necessary constructional details of a centre and a turning piece are shown in Fig. 106.

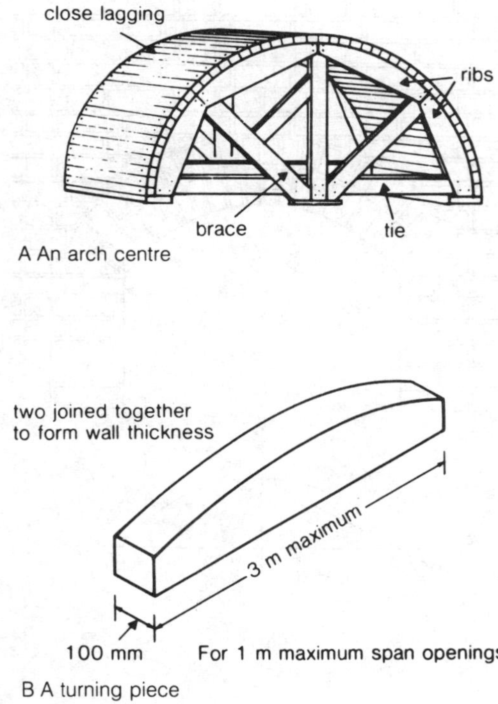

Fig. 106 Construction of a block arch

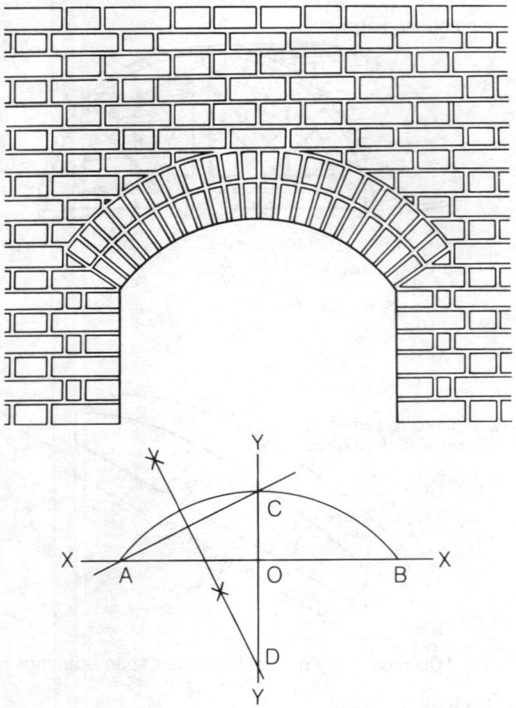

Method of construction
Draw lines X–X and Y–Y. Mark off distance O–C which is the rise and O–A–B, join points A and C. Bisect line AC and extend the bisector to line Y–Y at point D which is the striking point. Fix the point of the compasses at D, describe an arc through ACB which is the intrados.

Fig. 107 Segmental arch

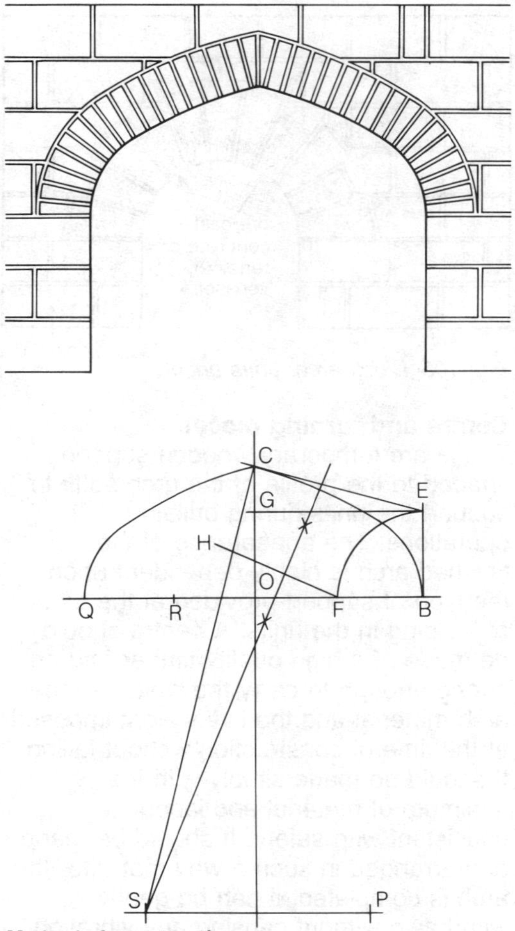

Method of construction
Draw the span AB and rise OC as given. Mark G 2/3OC and join GE with a line. Join CE with a line and draw a line at right angles to line CE. With the centre at B, describe an arc from E to cut AB at F. Mark CH = OG = BF and join HF with a line. Bisect HF and draw the bisector until it meets line CE at S. Mark off R and P and R, F, S and P are centres.

Fig. 108 Tudor arch

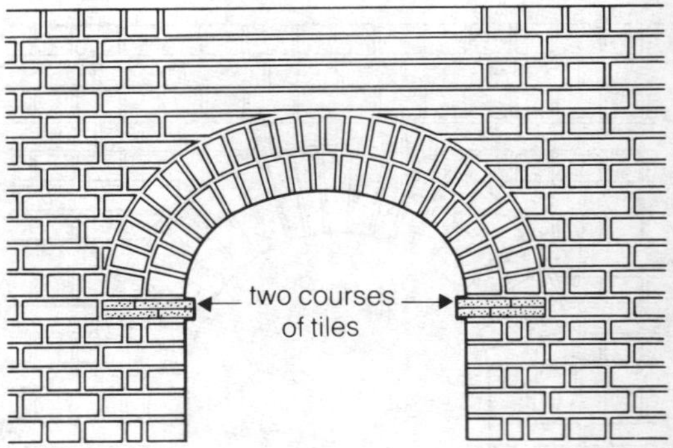

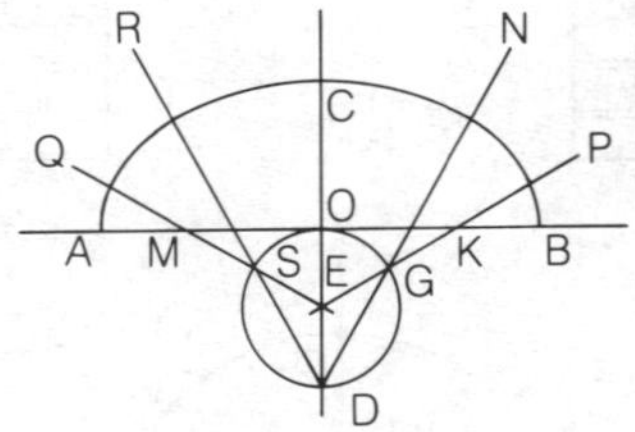

Method of construction
Draw span AB and rise OC = OD. With centre E, draw a circle touching O and D. QE, RD, ND and PE are common normals and the centres for the arch are M, S, D, G and K.

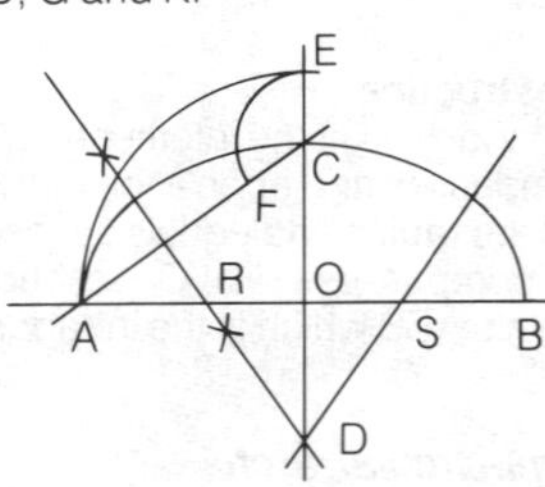

Method of construction
Draw span AB and rise OC. Join C and A with a line. With centre at O, describe an arc from A to E and, with centre at C, describe an arc to cut AC at F. Bisect AF and draw the bisector to cut AB at R and meet perpendicular line at D. Centres of the arch are D, R and S.

Fig. 109 Semi-elliptical arch

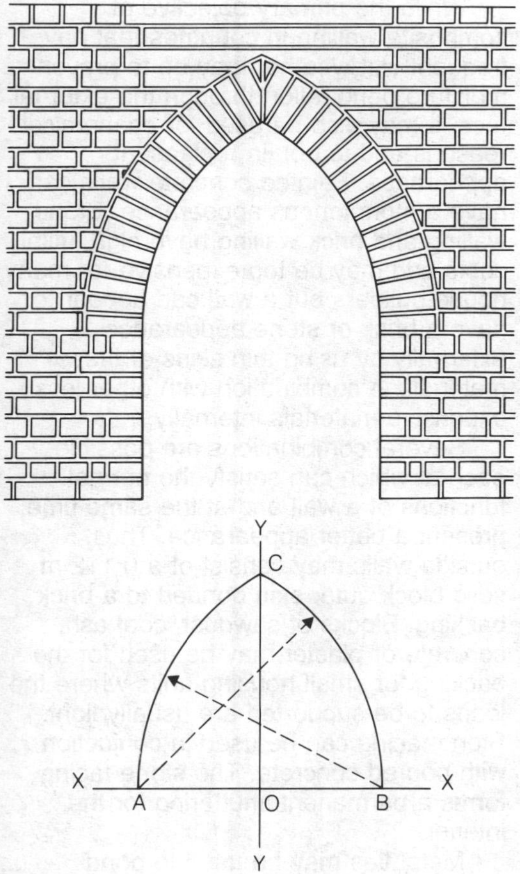

Fig. 110 Equilateral arch. A member of the group of Gothic or pointed arches which also includes the Lancet arch, drop arch and venetian arch. They all serve the same architectural purpose

Method of construction
Draw lines X–X and Y–Y. Mark off the span AB and rise OC, A and B are the centres. With the point of the compasses at centre A, describe an arc B to C and,changing the compasses to centre B, describe the second arc AC. The voussoirs are drawn as radials from the centres.

Cavity walling

A cavity wall consists of two thin walls with a continuous 50 mm air space in between them. The two walls come together only at openings and at the level of termination. Cavity walls are rarely used in tropical climates. The reason is clear. The main purpose of a cavity wall is to prevent the penetration of moisture through the external surface to the inner surface. The central air space in the wall functions to prevent such movement of moisture. The penetration of moisture through walls is not a major problem with buildings in tropical areas. The seasons are well defined in some tropical areas, so that there is no season of persistent wetness. Even in areas of persistent wetness, a heavy downpour is often followed by a spell of sunshine and warmth, which evaporates the water in the walls before it has time to penetrate through them.

To solve the problem of getting rid of excess heat within buildings in hot climates, cavity wall construction can be used to great advantage by helping to maintain a constant temperature. The inner wall has no direct contact with the hot outside air and, moreover, the layer of air between the two separate walls is a very poor conductor of heat, especially when the cavity is ventilated.

Where a cavity wall is to be ventilated, precautions should be taken to prevent the cavity becoming home for reptiles, birds and bats. Air blocks at the base of the wall should have fine wire netting placed over them before the external rendering is done, and eaves must have netted air inlet/outlet points. Figure 111 illustrates this practice.

Cavity wall ties

The two leaves of the cavity wall are tied together by building in non-corroding metal wall ties to bridge the air space. The ties, which should be staggered, are placed at pre-determined positions. A guide for spacing of the ties is to have ten ties for every 1·5 m^2 of walling. This does not allow for openings in the wall; ties must be placed close to reveals.

To keep the cavity clear of mortar droppings, a cavity lath, which is a piece of wood 50 mm wide by 25 mm thick and up to 1·6 m long with attached handles, is used, as shown in Fig. 111.

Composite walling

Composite walling, as the name implies, is a system of walling whereby the total thickness of the wall is made up of two different materials, one material for backing and another material for the facing. There are certain advantages to be gained by using this system of walling: a durable, weather-tight material can be used on the outside with a weaker and more porous material inside.

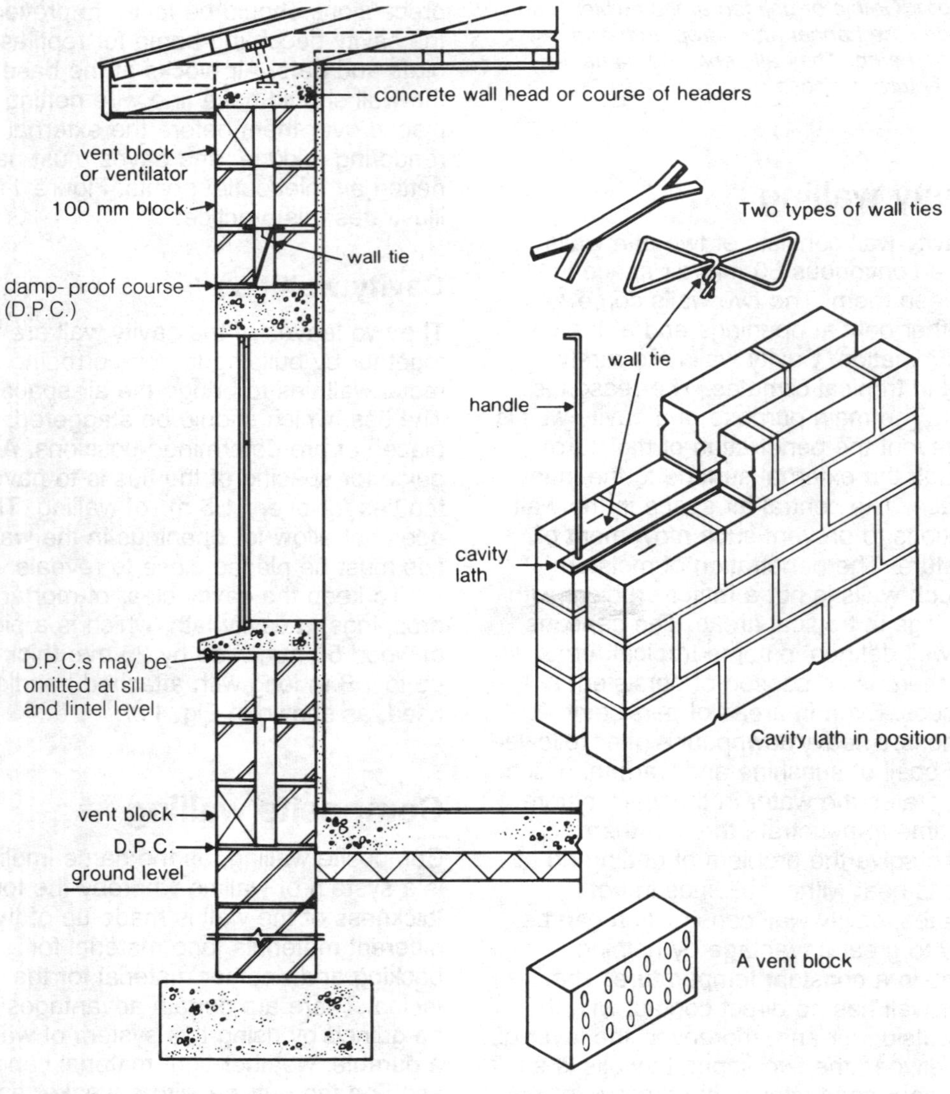

Fig. 111 Block cavity wall construction

While the primary objective of composite walling in countries that have frequent wet weather may be to prevent moisture penetration through the external face, it may also be used for economical reasons and to obtain a pleasant appearance. Painted concrete walls can have a monotonous appearance. Stone walling and brick walling have high initial costs and may be too expensive for many house builders but a wall can appear to have a brick or stone appearance externally by using thin skins of these materials in combination with other less expensive materials internally.

Several combinations are possible, each of which can satisfy the normal functions of a wall and at the same time present a better appearance. Thus, outside walls may consist of a 0·112 m solid block outer skin bonded to a brick backing. Blocks of sawdust, coal ash, concrete or plaster may be used for the backing for small housing units where the loads to be supported are usually light. Stone facing can be used in conjuction with poured concrete. The stone facing forms a permanent shuttering for the interior.

Metal ties may be used to bond together the two skins of composite walls. Some examples of composite walls are illustrated in Fig. 112.

Advantages of composite walling

1 Provides wider scope for choice of external wall finish.
2 Reduces the dead load of the wall.
3 Inferior material may be used for the

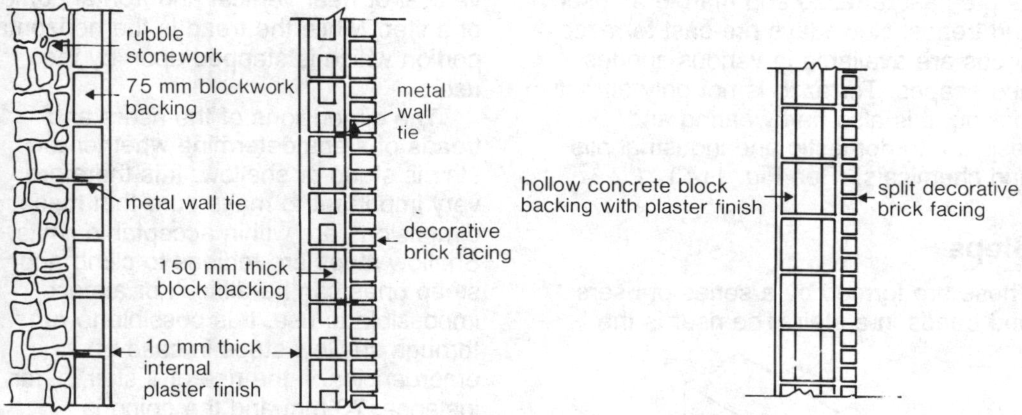

Fig. 112 Composite walls

inner skin, giving rise to a saving in cost.

4 Replacement of a skin can be effected without damage to the main wall.

5 Improves the weather-resisting qualities of the wall.

6 Can give a good standard of thermal and sound insulation.

Disadvantages of composite walling

1 More time may be spent in building the skins than building a single thickness wall.

2 The walls may become unnecessarily thick.

Fig. 113 Thresholds are sometimes used as a design feature to enhance the frontal appearance a building

Thresholds, steps and stairs

Thresholds

Sometimes the ground floor level of a building is made higher than the general ground level around the building as a design feature and sometimes the ground upon which a building is sited is such that the ground floor area is higher than the general ground level. In such cases, the entrance door to the building is provided with one or more steps, called a threshold, to facilitate easy access into the building.

The rules that affect the construction of stairs are also applicable to the construction of thresholds, except that the riser varies from 114 mm to 179 mm and the tread should be 279 mm minimum. Materials used for construction include stones, concrete and bricks or blocks. When bricks are used for thresholds, they must be of the engineering grade otherwise the edges will be damaged during constant use. Blocks are generally rendered with rich cement mortar.

Like stairs, steps in thresholds may be of the built-up type using materials such as pre-cast terrazzo and marble as risers and treads. Nowadays pre-cast terrazzo goods are available in various shades and shapes. Terrazzo is not only attractive looking, it is also hardwearing and resistant to domestic and industrial oils and chemicals. (See Fig. 114.)

Steps

These are formed by a series of risers and treads in a stair. The riser is the vertical or near vertical and frontal portion of a step, while the tread is the horizontal portion which is stepped upon by the user.

The dimensions of the risers and treads of steps determine whether the stair is steep or shallow. It is therefore very important to make sure that these dimensions are within acceptable limits. Shallow steps are tedious to climb and steep ones can be difficult or almost impossible to use. It is possible to run through shallow steps in case of emergencies. If the rise of a stair is, for instance, 70 mm and the going is 300 mm the stair will be shallow and if the rise is 250 mm and the going is 125 mm the stair will be steep. This is illustrated in Fig. 115.

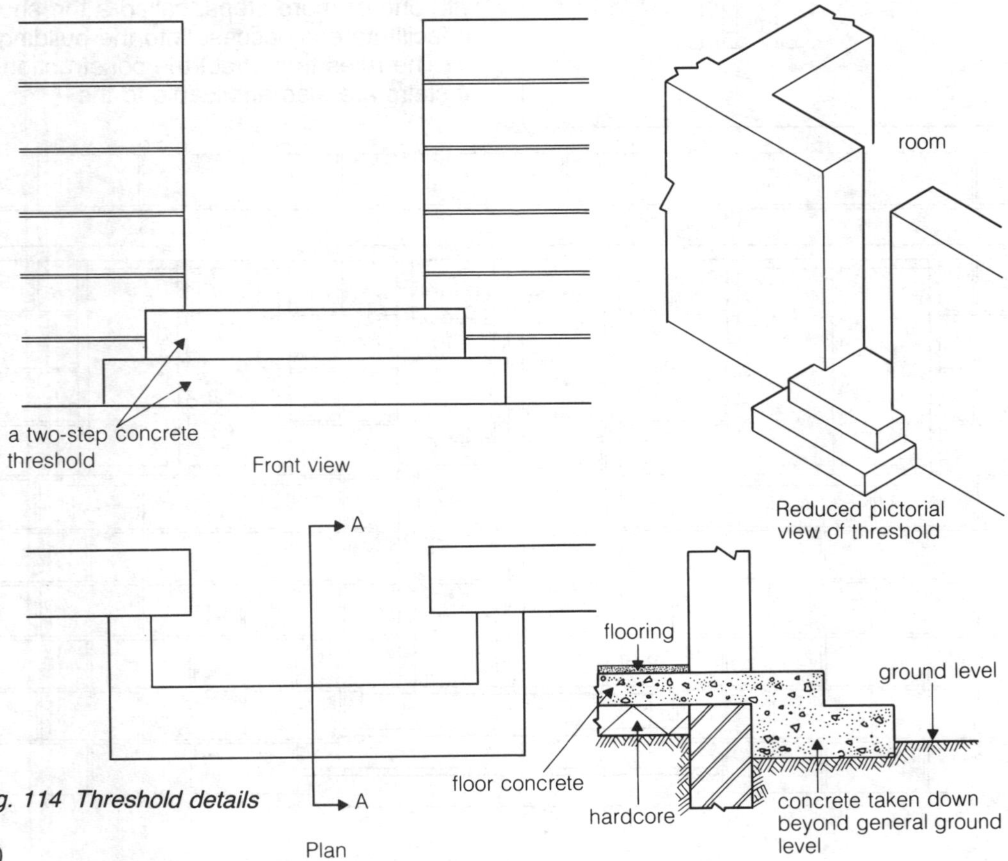

Fig. 114 Threshold details

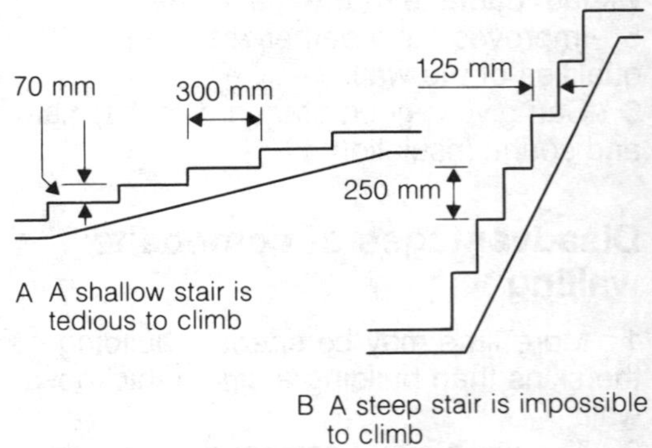

Fig. 115 Shallow and steep stairs

The riser and tread will vary according to the use to which the building will be put. By tradition, various dimensions of rise and tread have been accepted and these are given in Table 6.

Structure	Rise not more than (mm)	Going not less than (mm)
House	175–200	210–225
Block of flats	190	225–250
Factory	175–200	250
Public building	165	250
Theatre	150	275
School	140	275

Table 6 Ideal limits for stairs in various types of structures

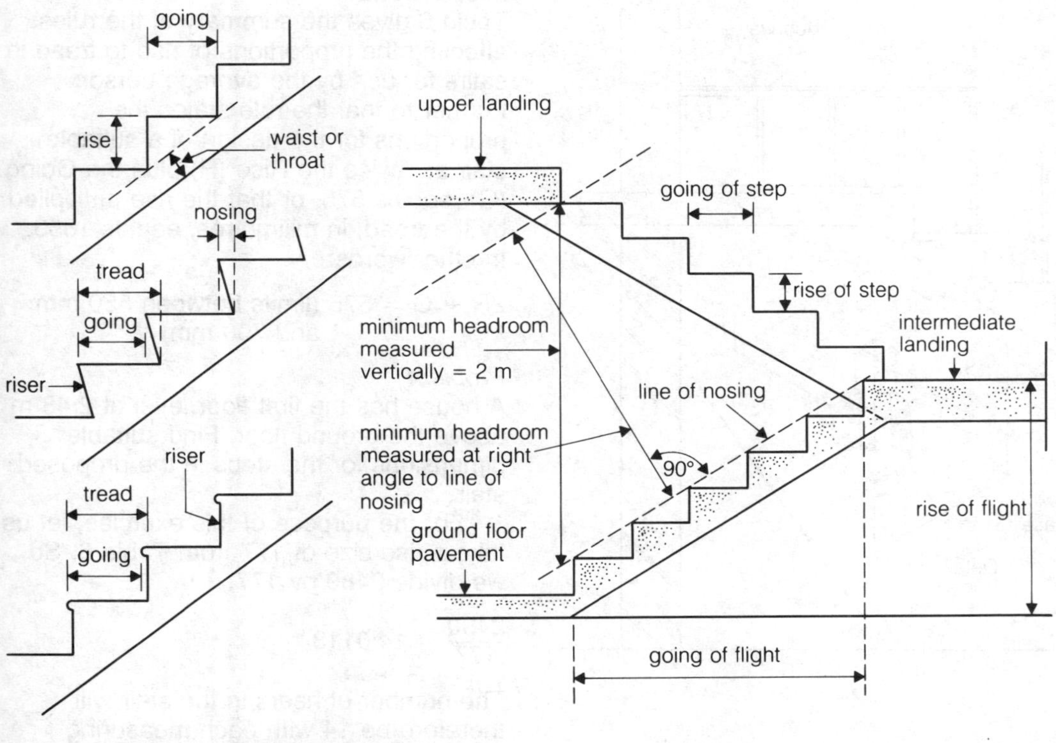

Fig. 116 Terms used in stair constructions

Terms used in stair construction

Staircase
This term most appropriately applies to wooden stairs because they have a case-like construction whereby the ends of the treads and risers are fixed into stringers on both sides forming a case containing steps in a flight.

Stairs
This term refers to a series of steps which facilitate walking access from one level or floor to another. Stairs may be constructed of wood, stone, metal and reinforced concrete. (See Fig. 116.)

Flight
This is an uninterrupted series of steps between two floors, or between a floor and a landing. (See Fig. 116.)

Landing
This is a level platform between two floors where the stair changes direction or to serve as a break in an extraordinarily long flight of steps. (See Fig. 116.)

Rise
This is the vertical distance between one tread and the next. It also refers to the vertical distance between the floor surface where the stair commences and the top surface of the next landing. When it is necessary to differentiate between these two meanings we describe the former as the 'rise of step' and the latter as the 'rise of flight'. (See Fig. 116.)

Tread
This is the width of one step. (See Fig. 116.)

Riser
This is the vertical or near vertical face of a step in a stair.

Going
This is the horizontal distance between the front face of one step and the next. (See Fig. 116.)

Nosing
This term refers to the horizontal projection of the tread over the rise. (See Fig. 116.)

Headroom
This is the permissible distance, measured either vertically or at right angles, from

the line of nosings of a stair to the landing of the next flight above or any similar obstruction. This distance allows people to move freely on the stair without hitting their heads against any obstruction. (See Fig. 116.)

Balustrade
This is a series of open-spaced supports for a handrail of a stair. (See Fig. 117.)

Balusters
These are small-section uprights connected to the treads of a stair to provide a guard and to support the handrail. (See Fig. 117.)

Quarter-turn landing
A quarter-turn landing is usually square, the same width as one flight of a stair. (See Fig. 118 A.)

Half-turn landing
A half-turn landing is equal in width to one flight and extends across two flights. (See Fig. 118 B.)

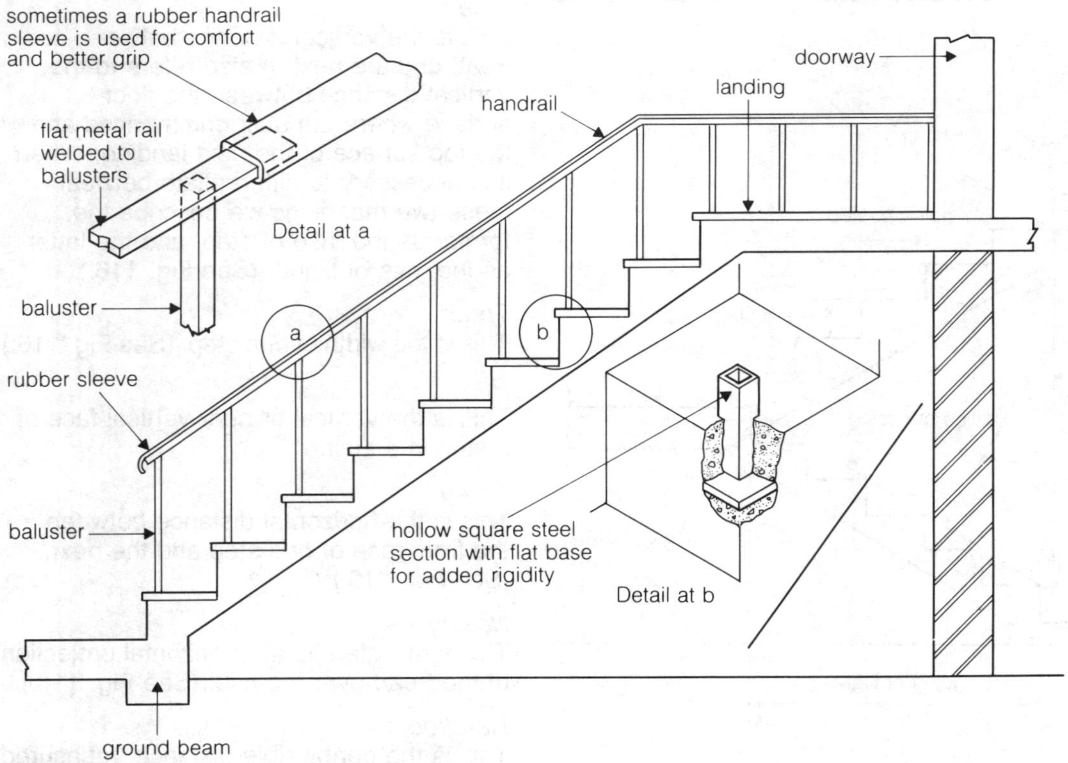

Fig. 117 A straight flight stair

Stairwell
This is an open space between flights of stairs. In a tall building, the stairwell can be a source of hazard unless it is adequately protected. (See Fig. 118 B.)

Fliers
These are the parallel steps forming a straight flight of stairs.

Rules and regulations affecting the construction of stairs

Proportions

Table 6 gives the summary of the rules affecting the proportions of rise to tread in stairs for use by the average person. Further to that the rule states the proportions for the design of a suitable stair as: twice the Rise (R) plus the Going (G), equals 575, or that the rise multiplied by the tread, in milimetres, equals 1650. In other words:

$$2R + G = 575 \text{ (limits between 550 mm and 700 mm)}$$

Problem
A house has the first floor level at 2·48 m above the ground floor. Find suitable dimensions for the steps in the proposed stair.

For the purpose of this exercise, let us take a rise size of 177 from Table 6. So we divide 2480 by 177:

$$\frac{2480}{177} = 14{\cdot}0113$$

The number of risers in the stair will therefore be 14 with each measuring 177 mm. This agrees with the ideal length of flight (see the right-hand column). The going can be easily calculated using the proportion 2R + G = 575 as given above.

575 − (2 x 177) = G
575 − 354 = 221

A rise of 177 mm and a going of 220 mm will give a comfortable stair in this case.

Design and construction
The Building Regulations (British) Part K (1985) has laid down requirements for the design and construction of stairs. There are also various regulations specific to the local area with requirements which satisfy local practices, needs and living habits. The ultimate objective of every regulation is to prevent unapproved practices and accidents from occuring and to ensure a considerable life-span of structures. It is important therefore that the reader is conversant with these regulations as they pertain to the construction of stairs.

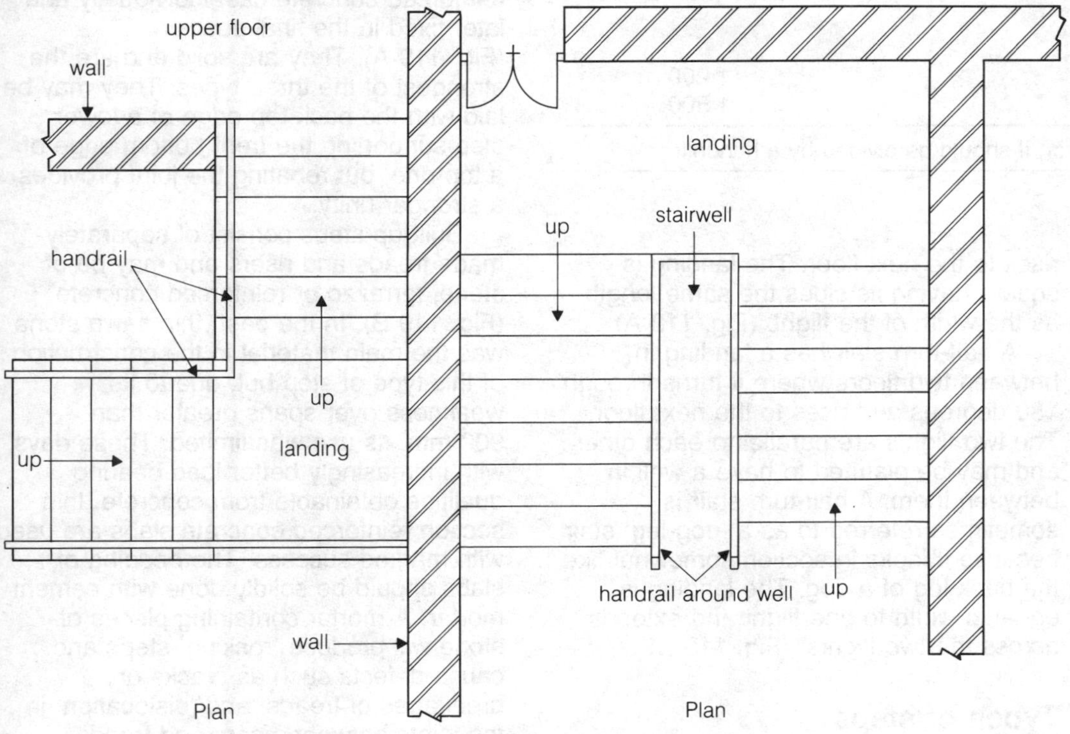

A Quarter-turn landing

B Half-turn landing

Fig. 118 Landings

Many accidents result from poorly constructed stairs. It often happens that stairs are drawn by architectural draughtsmen who have little knowledge of design principles and are built by craftsmen who have little knowledge about such works. Once a stair has been designed and built into the building, alteration works are difficult and costly.

Length of flight
By definition a flight has a minimum of three steps. A flight that is too long becomes tiresome for young and old people to climb. There are various guidelines for determining the length of a flight. There are some regulations that limit the flight to a vertical rise of 3·00 m; some others put the length of a flight at between 14 and 16 steps, after which there must be a turn or a landing before the next flight. It is also normal to limit the number of flights in a straight run to two, after which it is necessary to provide a run.

Headroom
The size selected for the rise influences the total headroom available. A sufficient rise is therefore essential in order to provide for adequate headroom. The vertical distance measured from the nosing to the next landing or any form of obstruction should not be less than 2·00 m and the distance from the nosing measured at right angles from the line of nosing should not be less than 1·50 m (Fig. 116). Where the stair is totally enclosed, the headroom distance needs to be increased to 2·10 m to avoid a feeling of oppression when using the stairway.

Structure	Minimum width of stairs (m)
Houses up to two storeys	0·850
Houses and flats over two storeys but not more than four storeys	0·900
Four-storeys and over	1·050
Public buildings – governed by estimated number of persons as follows	
200 persons per floor	1·050
200–400 persons per floor	1·350
Over 400 persons, every 100 additional	0·150 up to max. 2·700
Public entertainment buildings	
Less than 300 persons per floor	1·200
Over 300 persons per floor	1·500

Note: Where the width of a stair is 1·800 m and over, it should be divided by a handrill.

Table 7 Minimum widths for stairs

Types of stair

Stairs are classified according to their geometrical form. The three most common types are:

1 straight flight,
2 quarter-turn (landing or spacing),
3 half-turn (landing or spacing).

A straight flight stair connects between two floors with neither a landing nor a change of direction, hence the term straight flight. However, in some special circumstances, a straight flight stair may have a landing in the middle without a change in the direction, to afford a 'breathing space', for the user.

A quarter-turn stair has a landing mid-way to the next floor and from the landing turns through 90 degrees and rises to the next floor. The landing is square having its sides the same length as the width of the flight. (Fig. 118 A)

A half-turn stair has a landing in between two floors where it turns through 180 degrees and rises to the next floor. The two flights are parallel to each other and may be planned to have a well in between them. A half-turn stair is sometimes referred to as a 'dog-leg' stair because it looks in section somewhat like the back leg of a dog. The landing is equal in width to one flight and extends across the two flights. (Fig. 118 B)

Types of steps

The steps of a modern stair may be made of concrete in a monolithic form or be pre-cast individually and later fixed in the required position. When a stair is being pre-cast, it should be made upside-down and, when being lifted into position, the nosings must be adequately protected.

Steps are basically of three types:

1 rectangular steps,
2 built-up steps,
3 spandril (spandrel) steps.

Rectangular steps are made of reinforced concrete cast individually and later fixed in the final position. (Fig. 119 A). They are solid and are the strongest of the three types. They may be laid with the back top edge of a lower step supporting the front bottom edge of a top one, but rebating the joint provides a stronger unity.

Built-up steps consist of separately made treads and risers and may be of stone, terrazzo or reinforced concrete. (Fig. 119 B). In the past, thin sawn stone was the main material in the construction of this type of step but, due to its weakness over spans greater than 900 mm, its use was limited. These days, with increasingly better load-bearing qualities obtainable from concrete, thin section reinforced concrete slabs are used with marked success. The bedding of slabs should be solidly done with cement mortar. A mortar containing pieces of stone will produce 'rocking' steps and cause defects such as cracks or breakages of treads, and 'dislocation' in the joints between risers and treads.

Spandril (spandrel) (Fig. 119 C) steps are sometimes used in place of

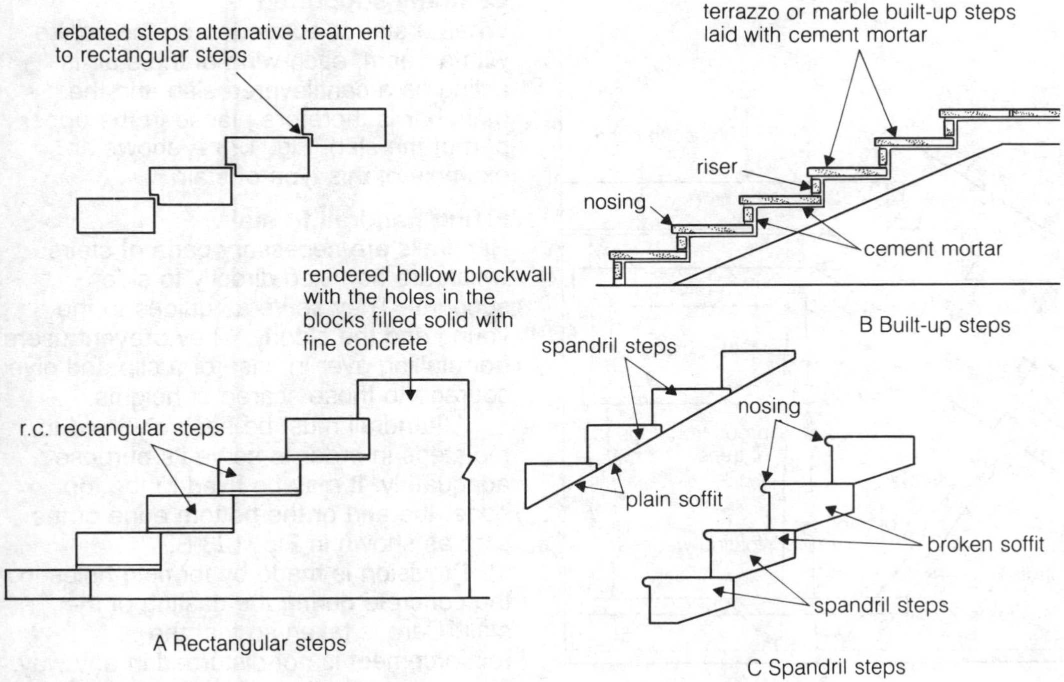

Fig. 119 Types of steps

rectangular steps because they have certain advantages over the rectangular type. Apart from the ends which are built into the wall, these steps are triangular in section. They are therefore lighter and provide a stair with a plain soffit, thereby increasing the space available for headroom. The use of spandril steps also enhances the appearance of a stair.

Reinforced concrete stairs

Of all materials used for building stairs, reinforced concrete is probably the most commonly used, hence this section is entirely devoted to the detailing of its construction. In doing this, the information about concrete covered in Chapter 10 is very important in understanding the reaction of concrete under different kinds of stress. As far as possible, locally made materials should be used for the construction of stairs. A formwork that uses logs for struts and props is shown in Fig. 120.

A reinforced concrete stair (Fig. 121) is no more than a simply supported slab or beam held on the slope. Its reinforcement is determined by how and where its supports are provided. The main reinforcement steel is placed in the direction of the span at the bottom. The distribution steel is placed above the main steel and at right angles to it. The exception to this detailing guide is when steps are cantilevered on a central support or supported only at one end. In this case, the principle of cantilevered beams applies and the reinforcement steel is placed at the top instead of the bottom. In instances where the ends of steps are built into side supports, top reinforcement steel is provided at the end or edges to prevent cracking at these points.

Straight flight supported at ends. The main steel is placed to span the supports as near as practicable to the soffit with sufficient concrete cover. In order to ensure that sufficient and uniform cover is provided for the steel, bar supports, which can be easily made on the site, are used. (Fig. 122 A)

Proprietary bar spacers made of cement or plastic are also available but the cost of these may be high. The distribution bars are then placed at regular intervals across the width of the stair on top of the main bar and without hooks. Stirrup wire binders are used to tie

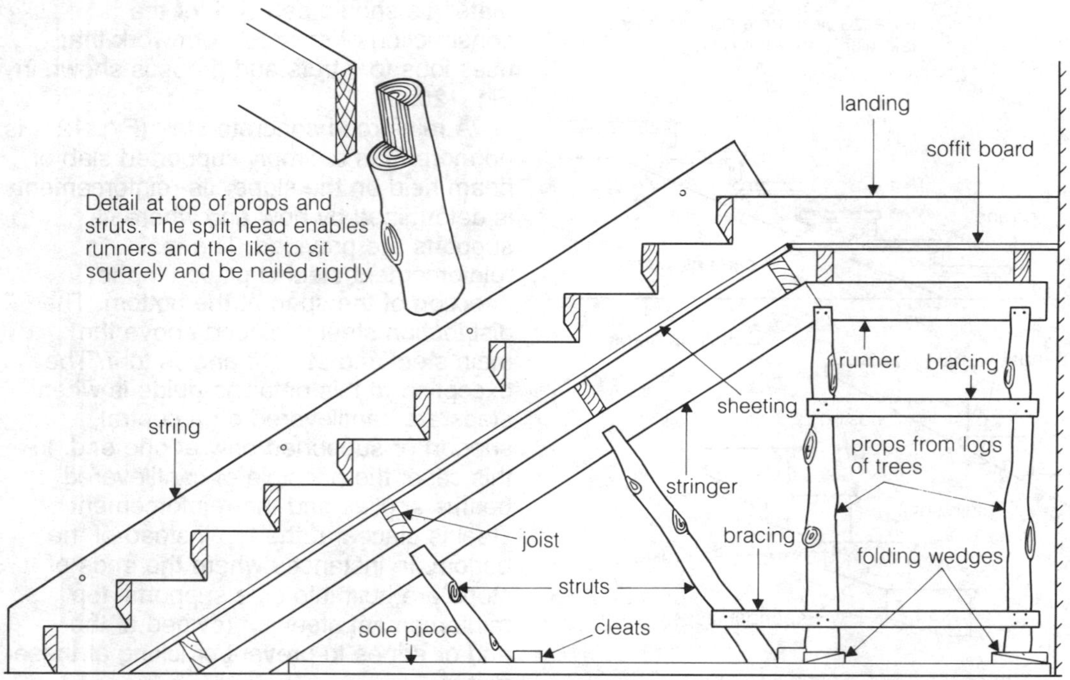

Fig. 120 A stair formwork

the two bars to ensure that the distribution bars remain at the correct positions until the concrete is poured into place. A common practice on the site is that of using the reinforcement already placed and tied in position as a means of access to the floor works above. This displaces the reinforcement and may adversely affect the performance of the finished work. Effort should therefore be made to cast the stair soon after the reinforcements are fixed.

Supports at sides
The method of reinforcement applied to this type of stair is exactly similar to that of a flat concrete slab supported on side beams. To provide hooks at the ends of the bars is optional, especially when the span is narrow. However, it is important to provide the special type of binders in the beam to prevent cracking on the upper edges. (Fig. 122 B).

Centrally supported
When a stair is supported on the centre with a beam, each wing of the stair is acting as a cantilevered slab and the main bar is therefore placed in the upper part of the slab. Fig. 123 A shows an example of this type of stair.

Fixing handrail to stair.
Handrails are necessary parts of stairs which are not fixed directly to side supports. They serve as guides to the young and the elderly. They prevent users from falling over in case of a slip and give courage to those scared of heights.

A handrail must be securely fixed to the steps in order to serve its purpose adequately. It may be fixed to the top edge, the end or the bottom edge of the step as shown in Fig. 123 B.

Provision is made by forming holes in the concrete during the casting of the stair. Care is taken so that the reinforcement is not disturbed in any way. The ends of the balusters are grouted in with cement at a later stage. Caulked lead is often specified for this fixing but the cost of lead may preclude its use for most works. Furthermore, cement grouting has proved quite successful, especially as most stairs are now formed of reinforced concrete.

Provision of non-slip surface
Slippery surface on treads and landings pose a great danger to users. Non-slip surfaces are therefore essential in concrete stairs especially when they are exposed to the rain. The following methods are generally applied to render treads and landings non-slippery.

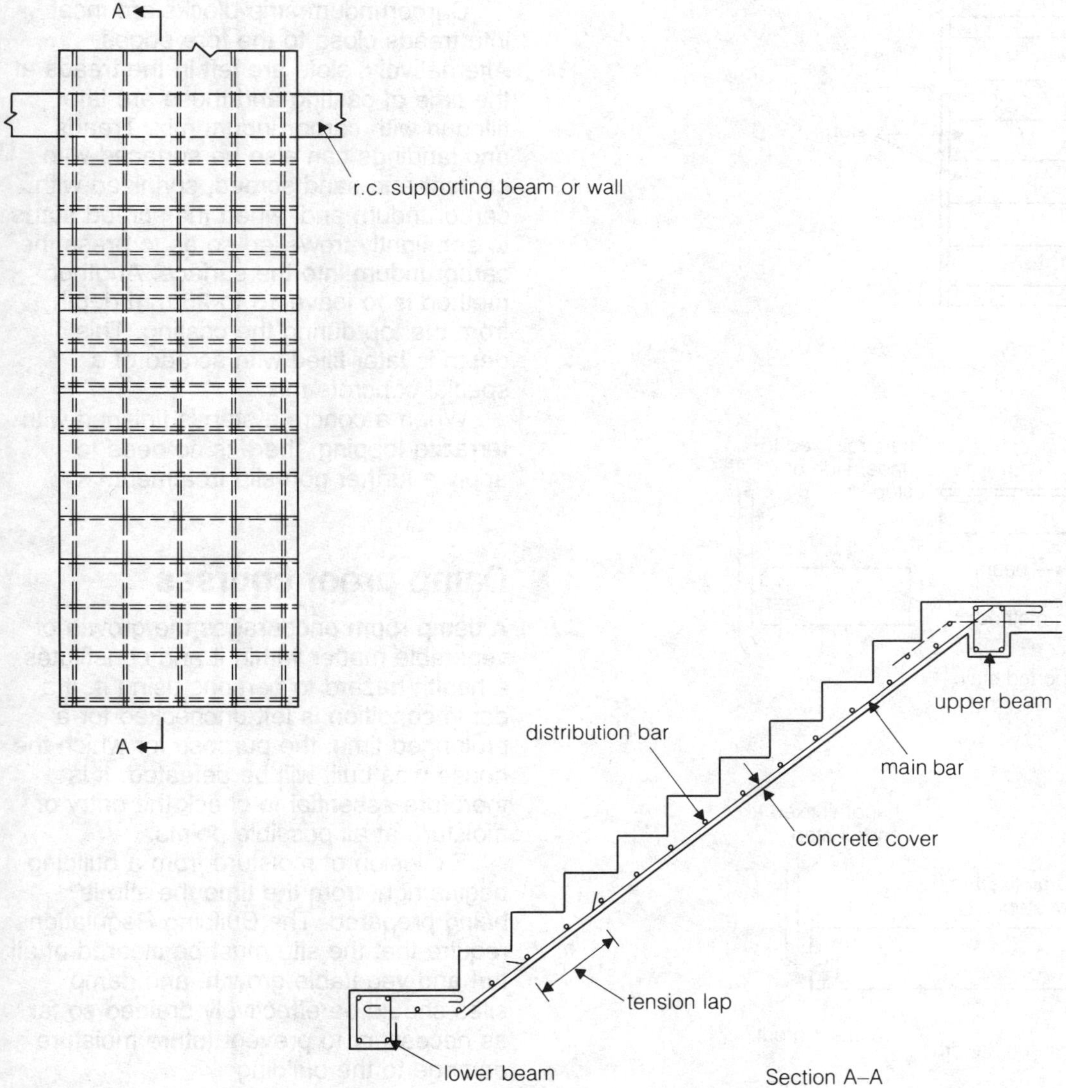

Fig. 121 Typical section through a stair

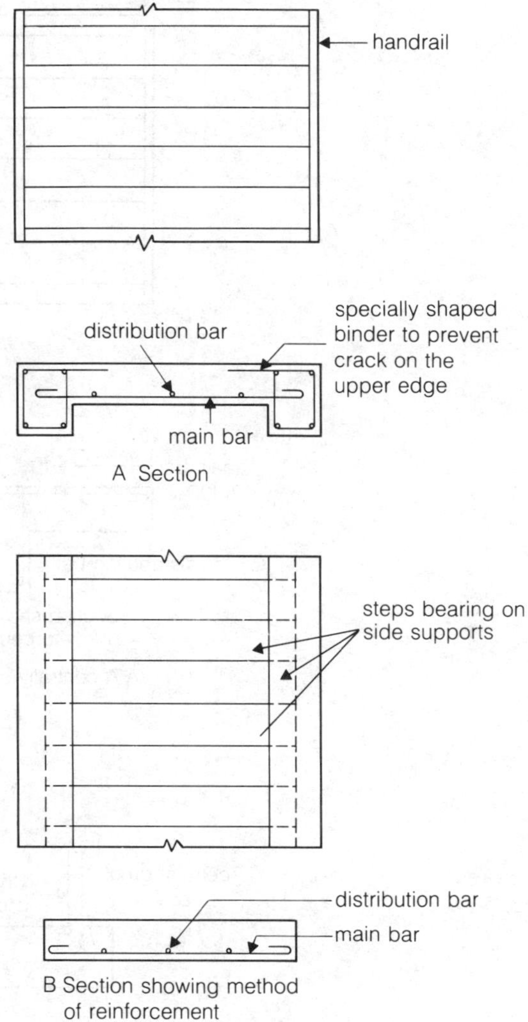

Fig. 122 End and side supports

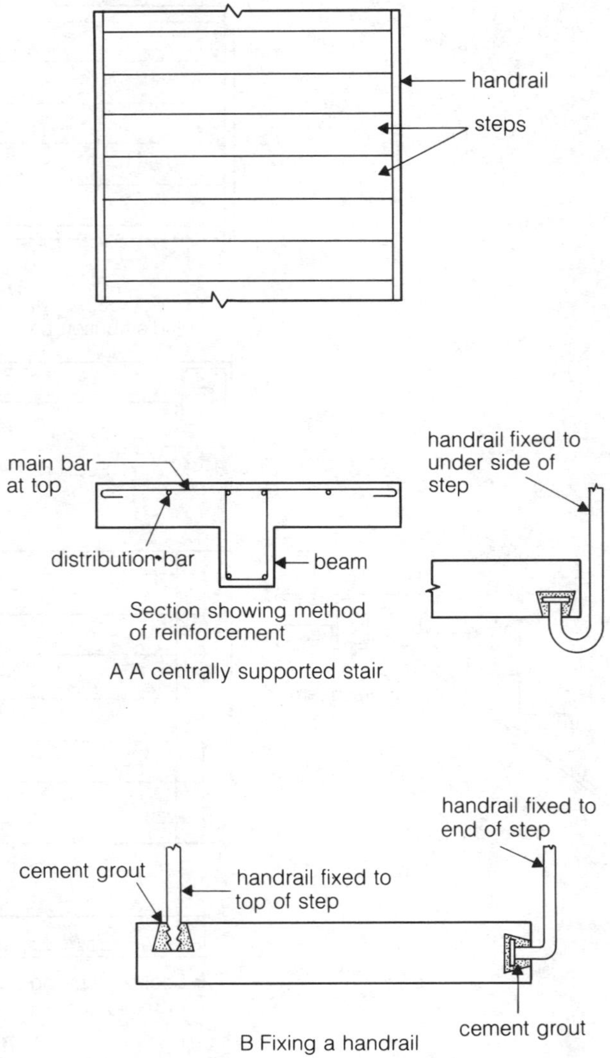

Fig. 123 Supporting a stair centrally and fixing a handrail

Carborundum strip blocks are inset into treads close to the fore edges. Alternatively, slots are left in the treads at the time of casting and these are later filled in with carborundum mix. Treads and landings can also be surfaced with cement and sand screed, sprinkled with carborundum and, when the screed starts to set, lightly trowelled so as to press the carborundum into the surface. Another method is to leave 13 to 25 mm depth from the top during the casting. This depth is later filled with screed of a special concrete mix.

When a concrete step is finished with terrazzo topping, there is no need to apply a further non-slip treatment.

Damp proof courses

A damp room encourages the growth of vegetable matter within it and constitutes a health hazard to persons using it. If a damp condition is left unchecked for a prolonged time, the purpose for which the house was built will be defeated. It is therefore essential to check the entry of moisture at all possible points.

Exclusion of moisture from a building begins right from the time the site is being prepared. The Building Regulations require that the site must be cleared of all turf and vegetable growth, and damp sites should be effectively drained so far as necessary to prevent future moisture damage to the building.

Dampness can gain entry into the building through the ground, through the external surface of the wall (as in driving rain) and through the top of the walls, floors and roofs.

Vertical damp proof courses (D.P.C.)

Vertical damp proof courses are required at positions vulnerable to horizontal penetration of moisture into the building. Places where vertical damp proof courses are used include:

1 reveals of door and window openings,
2 retaining walls,
3 basement walls adjacent to soil,
4 internal faces of walls enclosing laundry and dry cleaning undertakings.

The following materials are commonly used for vertical damp proof courses.

Mastic asphalt
This is supplied in blocks which, when heated, assume a plastic state which can be applied to wall surfaces. It is usually applied in two coats of 10 mm each to avoid leakage. It should conform to British Standard No. 1418 and can be used for all types of D.P.C. It is particularly useful in tanking to basements. Although it is very effective because it is jointless, to obtain the best results it has to be laid by specialists. When finished it has a glittery and smooth surface so it must be made rough while it is still hot, to form a good key for the subsequent mortar that will be applied on it.

Sheet lead
This is a rather expensive material as a D.P.C. Its flexibility, however, makes it an excellent material for vertical damp proof courses. It is used in cavity walls. A coating of bituminous paint on both sides is necessary to prevent the corrosion of lead which occurs when in contact with mortar.

Pitch polymer
This is supplied in rolls and can be used at the jambs of openings in cavity walls.

Horizontal damp proof courses

Horizontal damp proof courses must be provided close to ground level to check the upward movement of moisture from the ground. They should also be used below copings in parapet walls or at any other position where vertical movement of moisture in likely. Materials used include the following.

Bituminous felt
This has a base of hessian fibre or asbestos impregnated with two layers of bitumen. It is supplied in rolls and care must be taken to prevent tearing when unrolling the material. It is bedded on mortar, which must be free of stone or lumps capable of puncturing the felt. All joints should lap at least 100 mm. The Building Regulations require a horizontal D.P.C at 150 mm above the highest adjoining ground level.

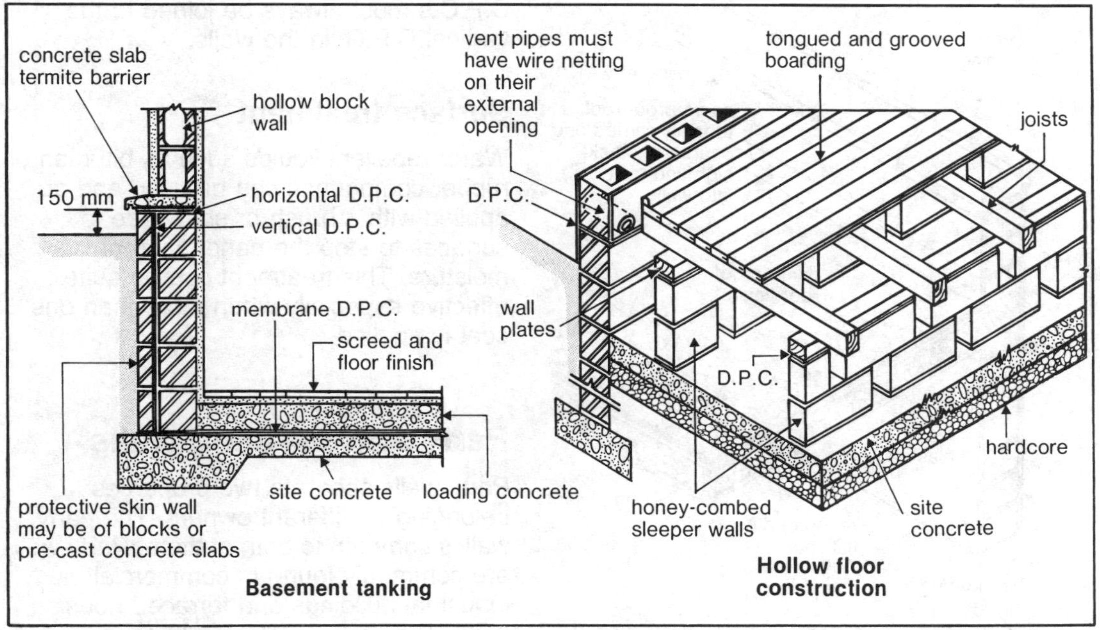

Fig. 124 Damp-proofing

Slate
This is a good damp proofing material when bedded in a 1·3 mm cement sand mortar. Slate must be laid in two courses and must be well lapped. It is a heavy material and may be expensive if slate is not readily available in your country.

Concrete slab
This is not an effective damp proof course under persistently wet conditions. However, in weather which is generally dry for most of the year, it can be used with a degree of success. When the walls reach about 150 mm above the highest part of the ground level, the floor concrete is laid on a consolidated bed of hardcore and made to extend over the walls. Figure 124 shows damp proof courses at various positions in the wall.

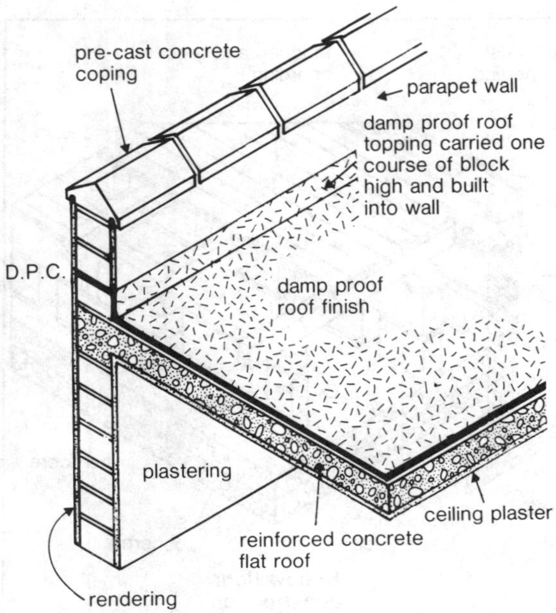

Fig. 125 Damp-proofing concrete flat roof

Membrane damp proof course

This refers to the incorporation of damp-resistant materials in a solid concrete floor. The membrane is laid on top of the structural concrete floor and the screed topping is then laid over it. Materials used include sheets of polyvinyl chloride, polythene, building paper and bitumen. Building paper is a layer of bitumen between two sheets of paper. Membrane D.P.C.s must always be joined to the vertical D.P.C. in the walls.

Surface treatment

Water repellent liquids such as bitumen or rubber compounds can be used and are applied with a brush or squeegee to surfaces to stop the penetration of moisture. This treatment can be quite effective especially when more than one coat is applied.

Party and partition walls

Party walls separate two properties belonging to different owners. The party wall is common to both of them. Party walls are commonly found in commercial or industrial buildings and terraced housing units. They are usually load-bearing.

Partition walls divide a floor space into rooms in a house. They may be constructed of a variety of materials, ranging from thin light sheets to heavy concrete blocks. This in turn means that they may be non-load-bearing structures. Space compels us here to limit our discussion only to block partitions. Blockwork partitions have replaced brickwork even where the external walls are of bricks, mainly for reasons of economy and reduction in weight.

Non-load-bearing block partitions

These types of partitions are usually constructed of hollow lightweight blocks. They are easy to build and relatively cheap. They are plastered both sides and fair faced blockwork is frequently used to good effect, especially when built in one of the stack bonds. However, partitions must always be plastered when built with concrete blocks because of the unattractive dull grey appearance of cement.

It is difficult to build a very thin, say 65 mm thick, partition wall. The wall is unstable during construction and may collapse under its own weight. It is necessary in such a case, to erect temporary profiles, which consist of 100 mm by 50 mm timbers with at least one straight edge, wedged into position between the floor and ceiling at the corners and stopped ends. On long walls intermediate profiles are necessary. The gauge for the work can be marked on the

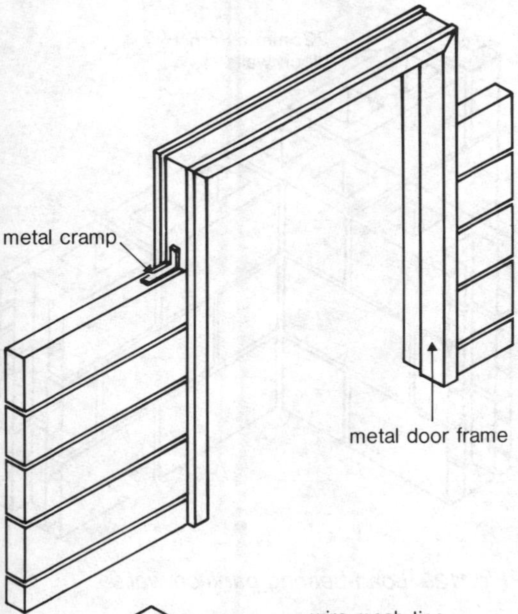

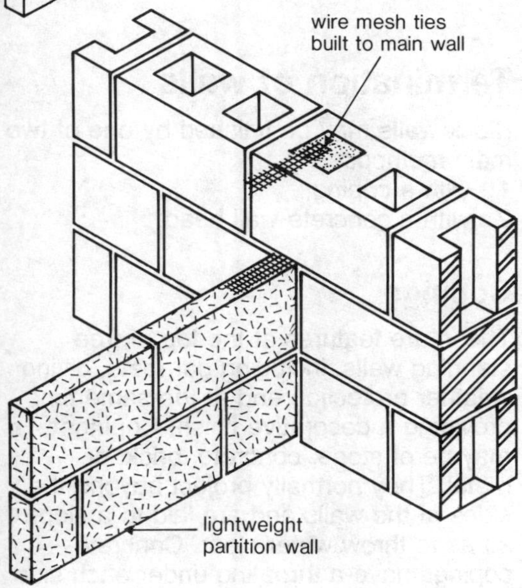

profiles. Door frames are measured and set up in position before the partition is built. The posts of frames are often extended and fixed securely to the ceiling structure. These aids help the blocklayer to carry out his work with reasonable safety, speed and economy. Such measures are not necessary if the partition wall is over 100 mm thick and not too high.

Lightweight block partitions may not bond into external walls. Instead, metal tie bars or mesh are used to tie the partition to the external walls. This is illustrated in Fig. 127. The bars or mesh must be built into the horizontal joints of the external walls, leaving the projecting end to tie into the partition. A case where this practice is possible is where the external walls are of fair face blockwork which is to be finished unplastered and the partition blocks are of a different material and finish. Cutting chases into the external walls into which the partition will tie may not only result in excessive waste due to breakages but also slow down the rate of working for the blocklayers. Figure 128 shows the construction of non-load-bearing partitions.

Fig. 126 Inserting door frame

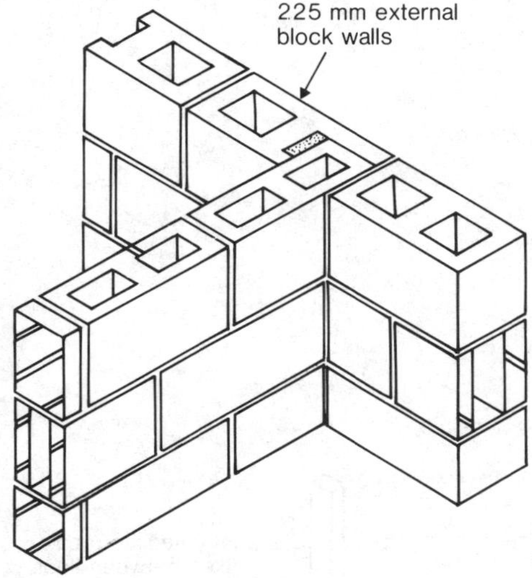

Fig. 127 Non-load-bearing partition wall

Load-bearing partitions

Most of these are constructed from the same size of block as external walls. The problems associated with the non-load-bearing types do not arise. Care should be taken at junctions so that the partitions are adequately bonded to the external walls. External walls are of 225 mm thick hollow concrete blocks and the partition walls are built with 150 mm thick hollow blocks. The depths for the two types of block are the same. (See Fig. 129.)

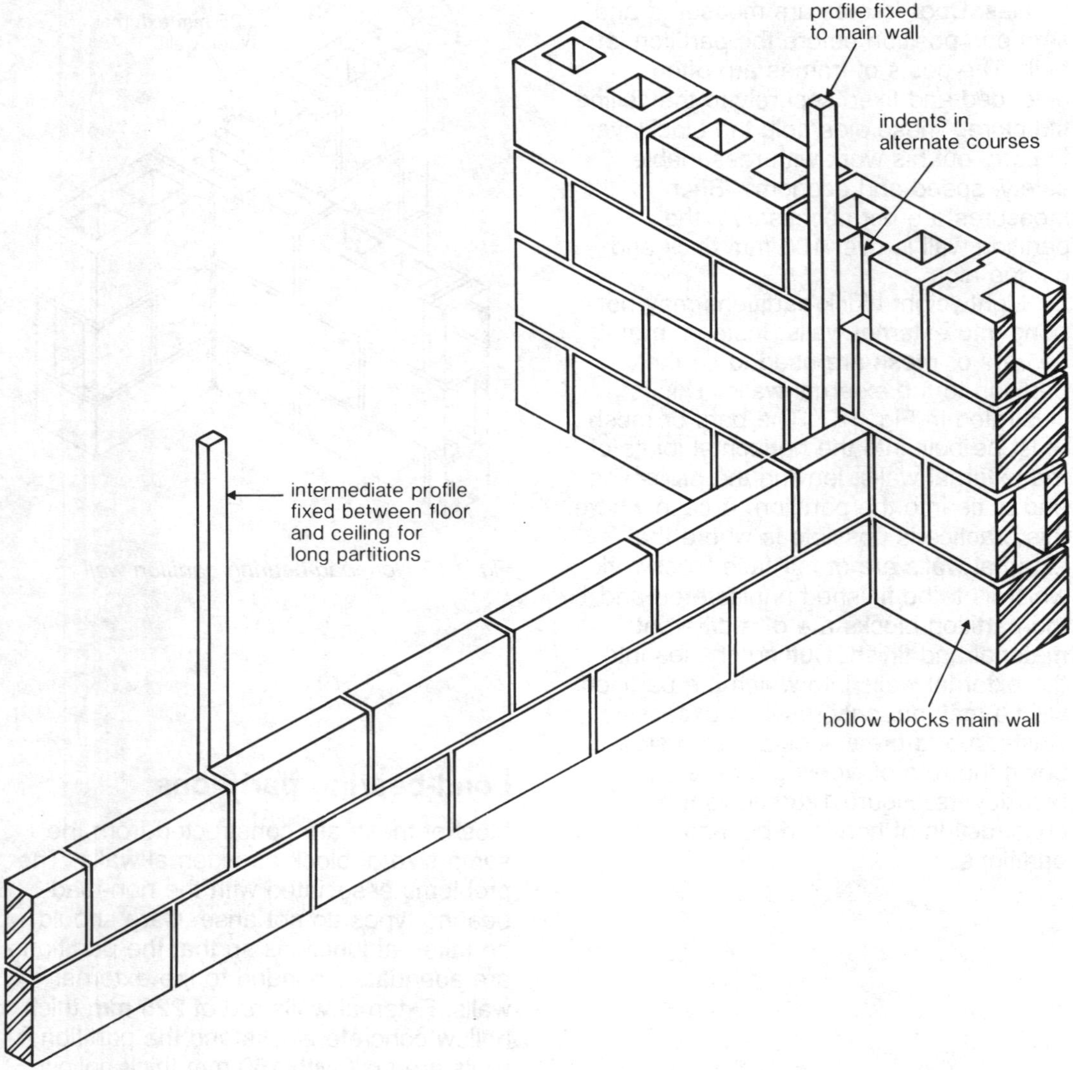

Fig. 128 Non-load-bearing partition wall

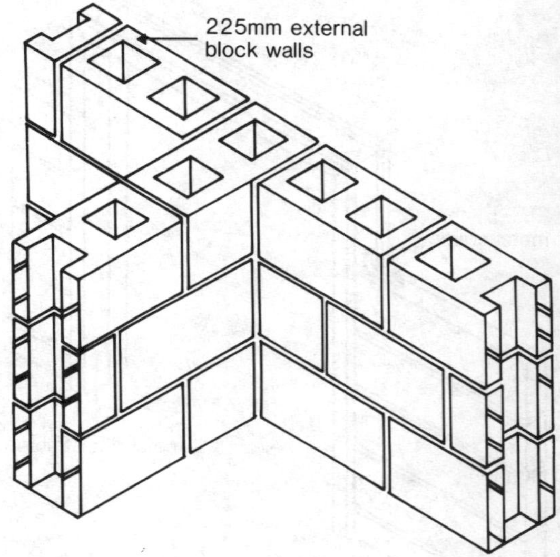

Fig. 129 Load-bearing partition walls

Termination of walls

Block walls may be finished by one of two main methods:

1 with a coping,
2 with a concrete wall head.

Copings

These are features at the top of free standing walls, including parapets, giving weather protection and at the same time providing a decorative finish. Copings may be of stone, concrete, brick, or metal. They normally project beyond the sides of the walls and are laid to a slope so as to throw water clear. Concrete copings have a throating under each side projection and a D.P.C. to prevent moisture from penetrating through the joints. They may be feather-edged, saddled-back, segmental or brick-on-edge. The different types are illustrated in Fig. 130.

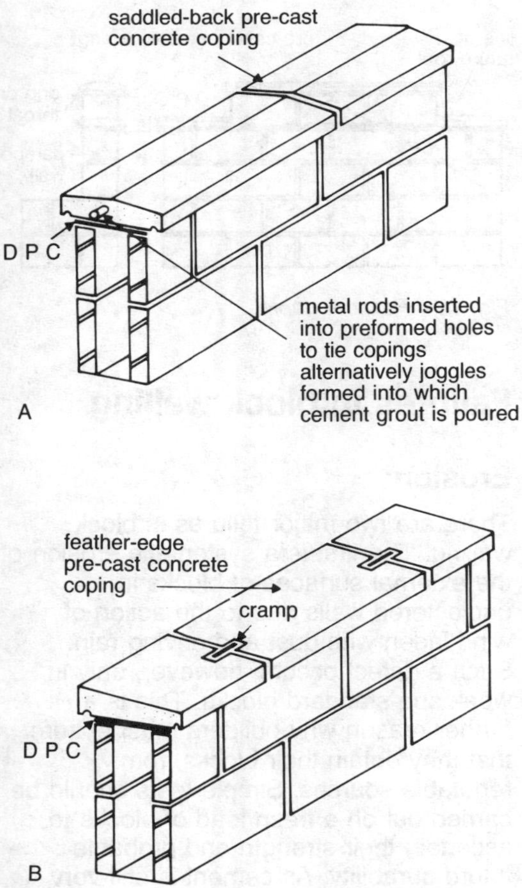

Fig. 130 Wall termination by copings

Concrete wall head

Concrete wall heads are found where mass concrete replaces the last course of blocks on the walls. The aim is to tie the roof rigidly to the walls, and distribute the load from the roof structure. Roof failures due to weak junctions are common. They are caused by wind suction lifting parts of the roof or the entire roof structure. Every year many houses have their roofs blown off completely during periods of high winds. These occurrences bring untold hardship to many families, who have to look for alternative shelters during the resultant reconstruction work. Thus a roof must be firmly anchored to its base and the coverings properly secured. A concrete wall head is illustrated in Fig. 131.

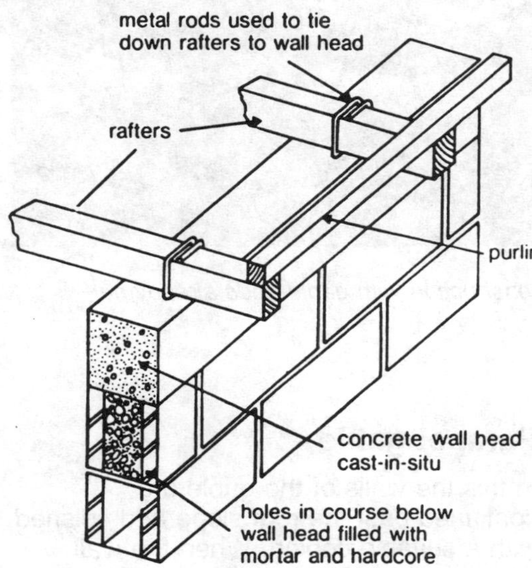

Fig. 131 Wall termination by wall heads

Gable walls

These are walls built to seal the ends of a pitched roof. There are two types of block gable:

1 verge gable,
2 parapet gable.

Verge gable

The gable is built to assume the same slope as the roof and terminate below the roof. The roof is continued over the gable and overlaps it, completely sealing the room below. This type of gable is easier to maintain as there are no extra precautions necessary at the points of contact with the roof. It results in a considerable saving in roofing materials compared with the roof being hipped. Figure 132 shows this type of gable. The wall is continued in the usual manner, the ends of the courses being racked back clear of the pitch line and the perpends from the main wall below being maintained

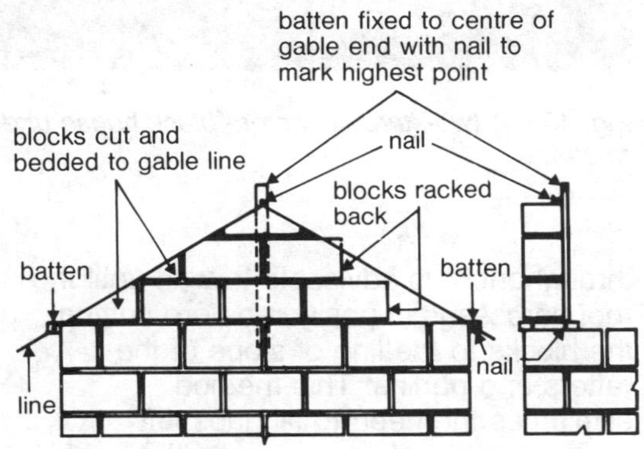

Fig. 132 Small verge gable

Fig. 133 A two-storey concrete block house under construction with gable ends already cut to shape

throughout. It is advisable to wait until the roof carcass is in position before cutting the blocks to the line of slope of the rafters and purlins. This method eliminates the need to fill gaps left between the roof covering and the gable slope which occurs otherwise.

Parapet gable

In this the walls of the gable are continued past the roof slope and finished with a suitable coping. When the wall passes the roof line, it is also called a parapet wall. The difficulty with this type of gable is the prevention of rain water slipping through the intersection of the roof covering and the inside of the block wall. A suitable means of sealing this joint must be employed. This is shown in Fig. 134 where the dotted lines show the slope of the roof.

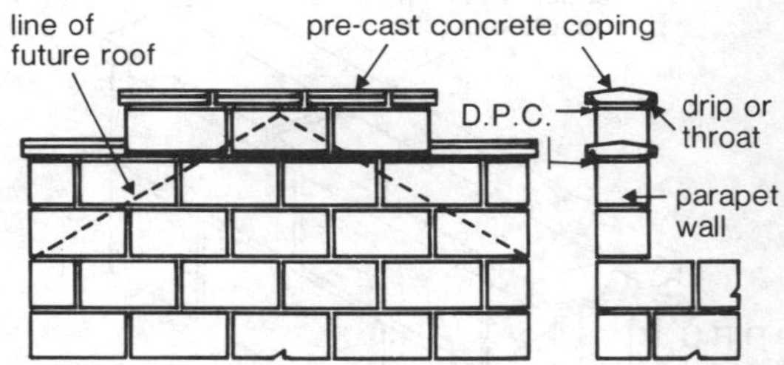

Fig. 134 Parapet gable

Failures in block walling

Erosion

There are two major failures in block walling. The first is a systematic erosion of the external surfaces of blocks in unplastered walls due to the action of wind laden with dust and driving rain. Such a defect occurs, however, only in weak sub-standard blocks. This is a further reason why builders must ensure that they obtain their blocks from reputable sources. Simple tests should be carried out on a fresh load of blocks to ascertain their strength and probable future durability. As cement is still very expensive private builders tend to economise on the use of cement to save money. Weak blocks will always, in the long run, turn out to be expensive due to the remedial work necessary to correct failures. Even when the wall is plastered, weak blocks provide a poor bond between themselves and the plaster, especially if the plaster is richer in cement than the blocks, as is often the case.

Cracking

The second common failure in block walling is cracking. This may not seriously affect the stability of the house structure, but cracks are unsightly and can be quite costly to remedy. Cracks may be due to several causes.

Shrinkage

Current research work on concrete products has established that drying shrinkage of concrete masonry walls varies with the moisture content of the units at the time of laying them. There are two categories of moisture movement.

1 Reversible movements caused by moisture changes in the fine particles of the material and the porosity of the blocks. Moisture movements due to fine particles are not normally serious unless there is an appreciable amount of silt clay or other similar material. It is to be noted that in no case will this type of movement take place unless moisture is present. Every effort should therefore be made to make walls watertight. Plastering the walls can greatly help in checking the horizontal entry of moisture into block walls.

2 Irreversible movements in cement products usually start during the setting of Portland cement in the form of shrinkage due to the formation of gels which shrink during drying out. A scientific explanation of this type of movement is unnecessary at this level. It is however, necessary to note that the presence of aggregate exerts a considerable internal restraint on the shrinkage of cement paste. The higher the aggregate to cement ratio the less will be the resultant shrinkage.

Shrinkage cracks in walls occur either vertically through the units and joints, or zig-zaggedly through the joints only, depending on the relative strengths of mortar and block and upon the bond between them. A significant proportion of the overall shrinkage of blockwork is due to the mortar joints. To limit cracking, which may be inevitable to some degree, the mortar must be made relatively weaker than the blocks. Weaker mortar also permits the blocks to move individually, thereby allowing a relaxation of stresses by redistributing the forces inside the wall. If cracks occur despite the use of weaker mortar, they are usually fine and are referred to as hair cracks which are centred around each block rather than being wide and more extensive.

Overloading

This may be the cause of cracks developing in a wall. Cracks due to overloading often originate from the loading points. Overloading may also result in a total collapse of the wall. To avoid the dangers of this type of failure, load-bearing block walls should comply with B.S. 5628: Part 3: 1985 or Part A Schedule 1 of the Building Regulations 1985.

Uneven settlement of foundations

A foundation may settle unevenly due to the action of traffic or the nature of the subsoil. The soil may contain weak spots and when these are systematically compressed by the weight of the building, settlement occurs at these spots. This can lead to cracking of the blocks. Cracks of this nature emerge from below ground level and the remedy is often costly involving massive temporary structures to hold up the wall while the foundation is replaced. Foundations should be carried down to a solid bearing stratum where they are unlikely to be disturbed by traffic vibrations. Most streets in urban areas are narrow but with heavy traffic and this causes an additional problem. The risk of cracking can be reduced by bonding and the use of mesh fabric in the horizontal joints.

Cracking in block walls may also be due to excessive droughts, especially if the foundation of the building comes upon a cohesive soil. The water contained in the soil evaporates under strong heat making the soil contract leaving gaps into which the foundation resettles. This causes cracks in the walls.

Impact

Cracks may be caused in slender block walls, especially internal ones, by the frequent slamming of doors. School hostels will be most prone to this type of cracking. It can be checked by fixing door springs to all doors so that they close without slamming.

Precautions to be taken against cracking

1 Adequately cure all blocks and ensure that they are dried enough before using.
2 Incorporate control or design joints at appropriate intervals and locations.
3 Introduce mesh reinforcement into horizontal joints if necessary.
4 Use cement mortar weaker than the blocks being used.
5 Make sure that the foundations are sufficiently deep.

Earth-retaining walls

Earth-retaining walls are required for a variety of purposes in building and civil engineering works. A basic knowledge of the construction of earth-retaining walls is essential, particularly for workers in developing countries, where professional advice may not be readily available or may be too costly for most people that need it.

Retaining walls may be required in the following circumstances:
1 to retain the lower portion of embankments and cuttings, such as in railway lines,
2 around the perimeter of levelled sites, such as sports fields,
3 wing walls to bridges,
4 basement walls to buildings,
5 water-retaining walls in reserviors and swimming pools,
6 grain-retaining walls in below-ground silage.

Fig. 135 House construction in Bangkok, Thailand

Materials

Retaining walls may be constructed of burnt clay bricks, sandcrete blocks, stone and concrete.

Bricks for retaining wall construction should be selected for their high permissible compressive stress and durability. It is advisable to consult the local brick manufacturers for advice on the selection of bricks in order to ensure that they are of low soluble salt content and are not excessively permeable.

Sandcrete blocks may be the filled or the solid type. The sandcrete block retaining walls depend mainly on the strength of the blocks and the stiffness of the wall. It is important, therefore, that the blocks are of the best grade. Blocks with an average compressive strength of 14 N/mm^2 and over are considered suitable.

Stone selected for earth-retaining wall construction should be hard, durable and strong. Rocks of the igneous family, of which granite is a member, are the most suitable. Igneous rocks are harder to work and more costly to use for masonry walls, but they are not as prone to chemical action and disruption as sedimentary and metamorphic stones. Chemical action and disruption is caused by soluble salts absorbed from the adjoining soil or, in industrial areas, the atmosphere.

Reinforced concrete is extensively used for the construction of high class retaining walls. Reinforced concrete retaining walls depend mostly on the quality of their design and construction for a satisfactory performance. The design of reinforced concrete retaining walls is outside the scope of this book. We shall therefore limit our discussion to the construction of masonry gravity earth-retaining walls as follows.

Unreinforced block and brick retaining walls

The earliest type of earth-retaining walls were constructed of bricks or stones. They depended on sheer self-weight to achieve stability and resist failure through overturning.

Brick retaining walls can be built either plain with vertical back, with attached piers or with buttresses, as illustrated in Fig. 136.

The self-weight of brick retaining walls should be sufficient to counteract the tensile stress caused by the lateral load of the soil and water. The relationship between the height of the wall and its thickness is also an important factor in the stability of gravity retaining walls.

However, for ordinary jobs, it is reasonable to assume that a brick retaining wall is capable of resisting a tensile stress of 0·42 N/mm^2, in the ultimate load condition, provided the mortar used is not weaker than 1:4 cement: sand mixture, or 1:½:4½ cement:lime:sand mixture. The use of plasticisers in mortar should be avoided for building retaining walls as the mortar is left with too many voids on setting. A similar value can be assumed for the shear resistance. The resistance to lateral pressure afforded by the tensile strength and by the self-weight of the wall should provide a factor of safety of 2 against overturning. A factor of between 1·5 and 1·75 is adequate for shear. The factor is obtained by dividing the total horizontal resistance by the total horizontal active force, which should be not less than 2. Where the total active horizontal force exceeds the total horizontal resistance the wall will fail by sliding.

As a general guide, Table 8 shows the recommended thicknesses and maximum heights of walls for average soil conditions where the backfill has a horizontal surface and does not carry superimposed loads.

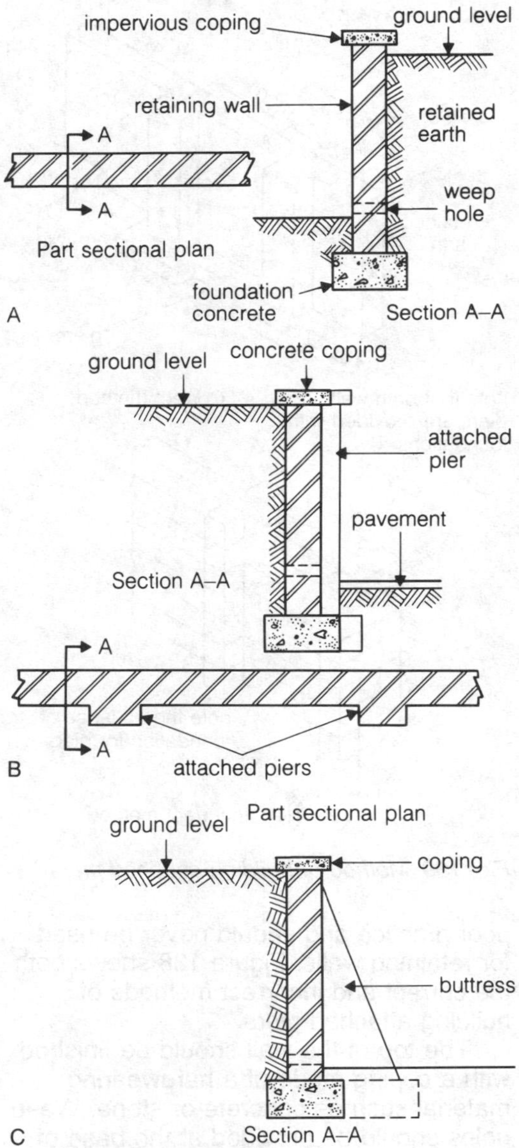

Fig. 136 Shapes of retaining walls

Thickness of wall (mm)	Maximum height (m)
552	1·80
450	1·55
225	1·45
200	1·30
150	1·10
100	0·80

Table 8 Recommended thicknesses and heights of block/brick retaining walls

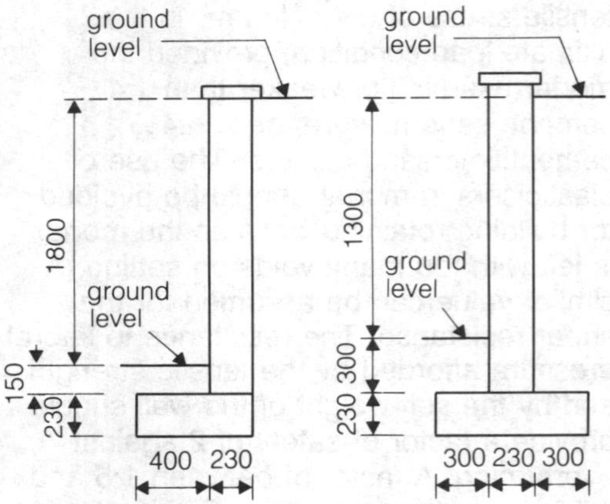

Fig. 137 Retaining wall foundations

The same general principles that govern concrete foundation for other brick and block walls also apply to retaining walls. The thickness of the foundation is not less than the thickness of the wall. The concrete bed may be laid flush with the back vertical face of the wall with a projection at the front of the wall not less than 300 mm, or it may have a projection of not less than 300 mm on either side of the wall. This is shown in Fig. 137.

Where attached piers are used, they should be spaced at intervals of 1·8 m centre to centre. Piers should have their width equal to the thickness of the wall and their length equal to the wall thickness. All piers are to rest solidly on concrete foundation. It is often seen that piers of boundary blockwalls are formed above ground level by turning a block across the wall thickness and then building subsequent courses on the block. This is

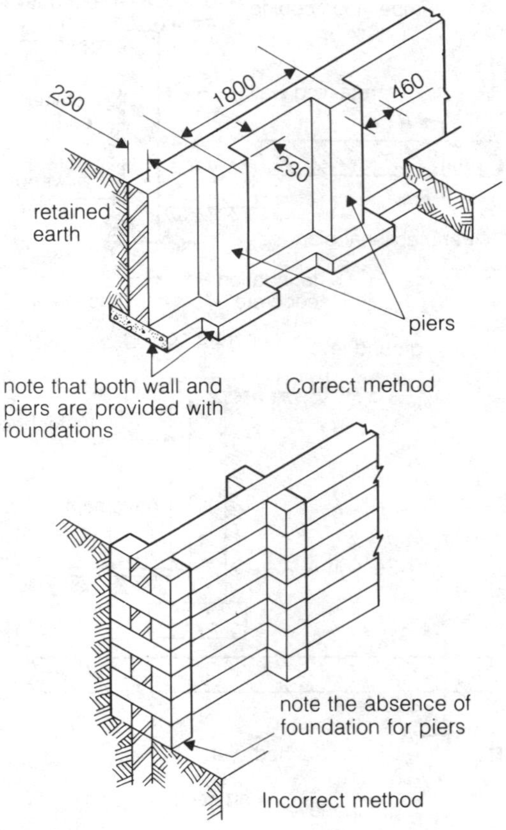

Fig. 138 Method of building attached piers

poor practice and should never be used for retaining walls. Figure 138 shows both the correct and incorrect methods of building attached piers.

The top of the wall should be finished with a coping made of a hardwearing material such as concrete or stone. Weep holes should be provided at the base of the wall to drain water from the retained soil. These holes may be formed by simply leaving every third perpend unfilled with mortar on the course at the ground or floor level as shown in Fig. 136. Alternatively, drain pipes may be used as shown in Fig. 139. To further encourage the drainage of water from the soil, the back of the wall may be packed with a layer of granular material, such as broken blocks, before the main backfilling.

Sandcrete block retaining walls are usually finished on the face with pointed joints. However, occasionally the back face may be rendered with cement mortar to make it waterproof.

Reinforced block retaining walls

Block retaining walls may be reinforced with steel in the same way as other reinforced block walls. A reinforced block retaining wall relies on the reinforcement

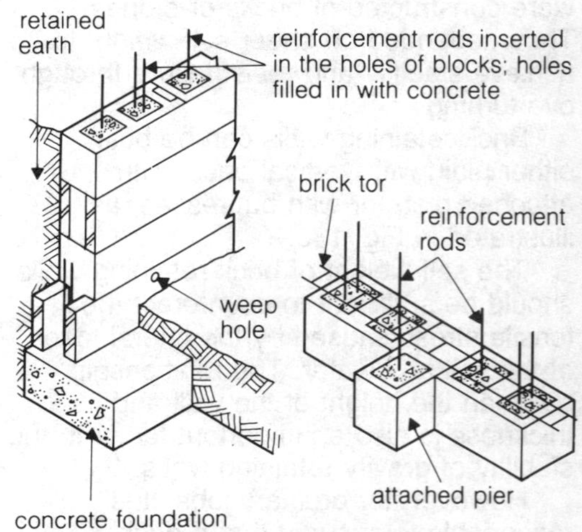

Fig. 139 Methods of reinforcing blockwalls

to resist the bending tensile stresses imposed by the backfill.

The blocks should be of the best type in terms of strength, materials and manufacturing process. The mortar should not be weaker than 1:4 cement:sand mixture or 1:½:4 cement:lime:sand mixture.

Walls should be left for at least ten days after the concrete is poured before backfilling is commenced. The reinforcements are introduced into the wall as shown in Fig. 139 and the holes filled with concrete. The filling should be done in lifts not higher than 1·6 m at a time in order to ensure that the holes are solidly filled and the reinforcements maintain their correct positions during the process.

Control joints are necessary in long walls. For unreinforced brick or block walls such joints are provided at intervals of 5 m, and in reinforced blockwalls they are provided at 8 m intervals. The joints may be plugged with an expanding material or expanded polyethelene and pointed with a weak gauge mortar of 1:3:12 cement:lime:mortar.

Stone retaining walls

A stone retaining wall relies on its self-weight to resist the bending tensile stresses imposed on it. Stone should be used for the construction of retaining walls not exceeding 1·45 m in height because of the necessity to increase the wall thickness, particularly at the base, as the height increases.

To prevent water pressure building up behind the wall, drainage pipes of clay, plastic or asbestos are inserted at about 1·5 m intervals at the base of the wall to drain off the water. A rubble filling behind the wall can be used as earlier stated to increase the drainage of water. Figure 140 shows a typical stone retaining wall construction.

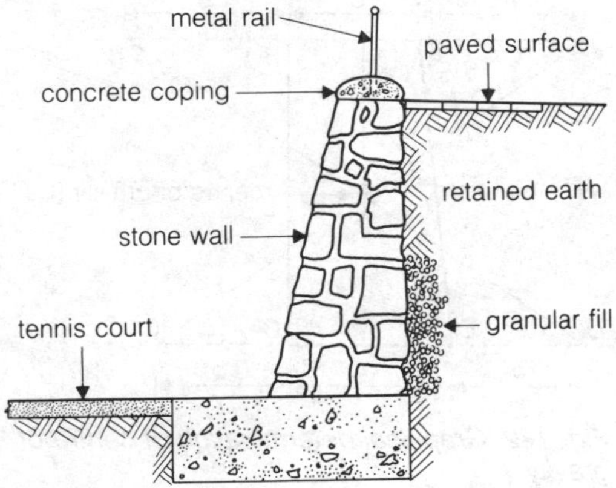

Fig. 140 A typical short stone retaining wall

Stability of retaining walls

Horizontal pressures of the retained soil are the most important consideration when dealing with retaining walls. These pressures will differ considerably according to the type of soil and the moisture content. If the soil is non-cohesive or granular, for example sand or gravel, its behaviour will, when retained in some manner, vary extensively from cohesive soils.

For a small demonstration of the behaviour of granular soils, consider the illustration in Fig. 141. A box having a sliding wall is filled with sand and then the sliding wall is lifted suddenly. The sand will begin to crumble down until a state of equilibrum is reached. The slide is along a shear plane and forms a slope. The angle which the surface slope makes with the horizontal is known as the angle of repose. This angle may also be called the angle of friction. For, in the case of retained water in which no friction exists between it and the retaining structure, the angle of repose is zero.

Different soils exhibit widely varying angles of repose. Moisture content of the material is also a significant factor in determining the angle of repose. If the moisture content of the sand in the demonstration is at an optimum value, the sand might stand vertically for a short time.

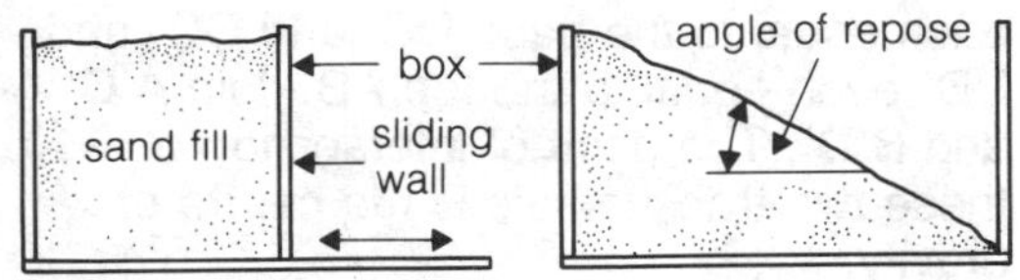

Fig. 141 Angle of repose of sand

If a retaining wall is absolutely rigid, earth pressure at rest will develop. If the wall deflects by moving slightly away from the soil, active earth pressure develops and if the wall moves towards the earth, passive pressure will develop.

Under normal conditions, earth pressure at rest is so intense that the wall deflects, relieving itself of the pressure, and active pressure results. For this reason, most retaining walls are designed for active pressure due to the retained soil.

The design of retaining walls involves the application of various earth pressure theories, which do not come within the level of discussion in this book. However, the middle-third law will be explained below. This law is used in the design of retaining walls.

Middle-third law

Consider the forces acting on a metre run of a retaining wall shown in Fig. 143. The forces acting are the horizontal thrust, *T*, due to the material retained acting at a centre of pressure, P, and the weight of a metre run of retaining wall, *W*, acting through the centre of gravity, C.G.

The centre of gravity is conveniently found using the graphical method as shown in Fig. 142. Draw AA′ and BB′ each equal to the base DC, and CC′ and DD′ each equal to the top AB. Join A′C′ and B′D′. The point of intersection of these two transversals is the centre of gravity.

The line of action of the resultant soil thrust cuts the vertical line, drawn through the centre of gravity of the wall section, at the point O (Fig. 143). From O a vector line is drawn horizontally to the left in multiples of the thrust. From O vertically downwards a vector line is drawn to represent the weight of the wall. The parallelogram of forces is completed and the resultant, *R*, of the two forces (diagonal) obtained, in a unit of force. If the resultant, *R*, falls outside the base line CD, the wall would overturn about the point D. If it cuts the base line at a point near D there would still be a tendency to tilt, thus producing tension in the masonry wall. To minimise this tension, the wall is so designed that the resultant force, *R*, cuts the base line between points K and L, known as the middle-third.

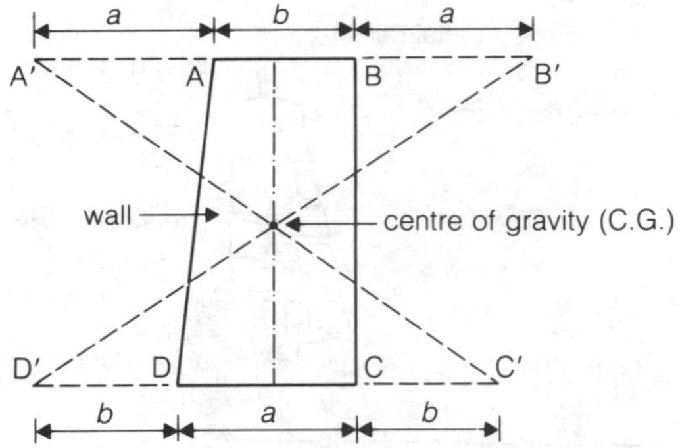

Fig. 142 Graphical determination of centre of gravity

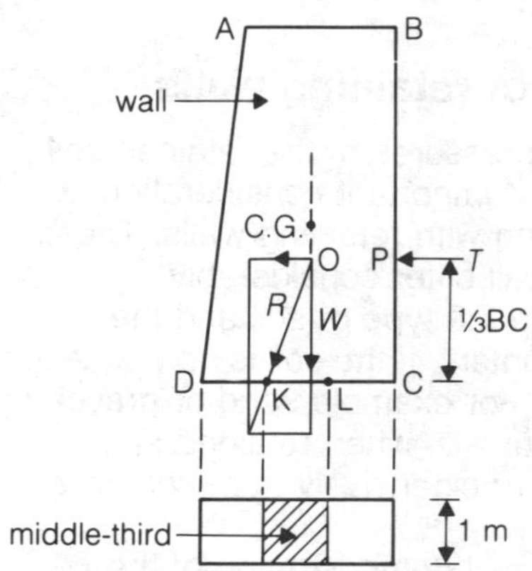

Fig. 143 The middle-third law

The middle-third law, therefore, states that when two rectangular surfaces are in contact the resultant thrust must not act outside the middle-third of the width of contact of the surfaces if tension is to be avoided.

It is important to note that tension in a masonry work is undersirable, as masonry is weak in tension.

Points to be observed when constructing masonry retaining walls

1 Horizontal damp proof courses should not be used in retaining wall constructions.
2 Concrete footings or foundations should be properly brushed down with a wire brush and cement water before the first course of wall is laid.
3 Long walls should be provided with control joints at pre-determined intervals and plugged with appropriate expanding material to accommodate the effect of expansion and contraction of the wall.
4 When porous blocks are used or the backfill material is impermeable, the back of the wall should be rendered to make it waterproof.
5 There should be no backfilling until the wall is sufficiently set, and in any case ten days should be allowed for the drying out of the wall.

Possible failures and remedies

Masonry earth-retaining walls may fail in the following ways.

Overturning

Where a wall and its foundation have insufficient weight to counteract the thrust from horizontal forces about the toe of the wall, it will fail by overturning.

To provide an ample factor of safety against overturning, the resultant thrust from the horizontal and vertical loads must fall within the middle-third portion of the base (Fig. 144 A).

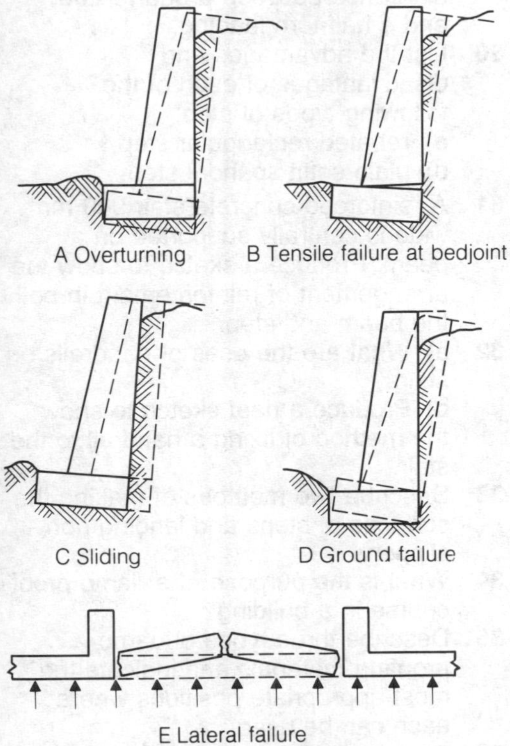

Fig. 144 Possible failures in masonry retaining walls

Tensile failure at bedjoint

Where a wall has insufficient weight, and very high tensile stresses are trying to develop due to the overturning moment, the wall will fail in tension and overturn (Fig. 144 B).

The thickness of the wall must therefore be increased to give sufficient weight to counteract the overturning moment from the horizontal forces in order to maintain compressive stresses across the whole of the bedface.

Sliding

Where the total active horizontal force exceeds the total horizontal resistance, the wall will slide (Fig. 144 C).

The factor obtained in dividing the total horizontal resistance by the total horizontal active force should be not less than 2, as previously stated.

Ground pressure

Where the actual ground pressure below the base of the wall exceeds the permissible ground pressure, the wall will fail (Fig. 144 D).

In certain cases where the permissible ground pressures are very low it may be advisable to use a reinforced concrete wall instead.

Lateral failure

This will occur in a buttressed wall when the panel between the buttresses is not thick enough to resist the tensile bending stresses as a result of horizontal loads (Fig. 144 E). This failure normally occurs at the perpends of the wall.

Revision questions

1 What are the advantages of laying out the materials properly before commencing a building operation?

2 Give five precautions to be taken in order to minimise the wastage of building blocks during a building operation.

3 To a height of four courses, show the face bonds of the following types of blockwork:
a) running or half bond,
b) stack header bond,
c) ashlar bond,
d) basket weave bond.

4 Briefly describe the purpose of screen walls. With the aid of a sketch, give an example of a screen wall.

5 Describe a method of setting out the first course of a block wall.

6 Using sketches wherever necessary, explain the following building terms:
a) ranging line,
b) profile,
c) racking back,
d) toothing,
e) datum level.

7 What are the three most important considerations to be borne in mind when building the corners of a wall?

8 a) List three methods of levelling and describe one of them in detail.
b) Define the following building terms:
i) eye-line,
ii) levelling pegs,
iii) boning rod.

9 Briefly describe the method of

building pre-cast concrete sills into the window openings of a long wall.

10 With the aid of clean sketches, show how corner blocks are used when building a long wall.

11 a) Describe the working procedure for fixing a window frame into a wall as the work proceeds.
b) Draw an isometric view of a fixing lug.

12 With the aid of suitable sketches, show two methods of spanning an opening in a block wall.

13 Draw a cross-section through a concrete lintel, indicating the concrete cover and the positions of the reinforcing rods.

14 Describe a working procedure for pre-casting a reinforced concrete lintel.

15 What are the advantages and the disadvantages of pre-casting concrete lintels?

16 What are the recommended minimum times for striking, at a normal temperature, the sides and soffits of the following types of concrete lintel:
a) a short span lintel not exceeding 1·6 m long
b) a long span lintel exceeding 1·6 m long,

17 With the aid of annotated sketches, draw the face of a brick arch over an opening and indicate the following:
a) span,
b) rise,
c) abutment,
d) reveal,
e) key brick,
f) voussoirs,
g) extrados,
h) intrados,
i) crown.

18 A short span concrete lintel is to be cast-in-situ. Produce a cross-sectional drawing to show the necessary formwork.

19 With the aid of annotated sketches, illustrate the following:
a) crank at the end of a reinforcing bar,
b) a turning piece.

20 Produce sketches to show the difference between a cavity wall and a composite wall.

21 What special precautions are necessary when ventilating a cavity wall in the tropics?

22 Produce neat sketches of the following and state in each case the purpose:
a) cavity wall tie,
b) cavity lath,
c) vent block.

23 What are the advantages of composite walling?

24 Produce neat sketches to show how the overlap in open riser pre-cast steps is obtained during fixing.

25 Use sketches to show the main differences between a threshold and a stair.

26 A block of flats has a vertical distance between two floors of 3 m. Calculate the dimensions for the going and rise for the steps.

27 Define the following terms as applied to stairs:
a) flight,
b) going,
c) headroom,
d) stairwell,
e) flier.

28 Give the permissible widths for stairs in the following types of structures:
a) houses up to two storeys in height,
b) houses four storeys and over,
c) public buildings with 200 persons per floor,
d) public entertainment buildings with less than 300 persons per floor.

29 With the aid of sketches, show the difference between a quarter-turn and a half-turn landing.

30 List the advantages and disadvantages of each of the following types of step:
a) rebated rectangular step,
b) plain soffit spandril step.

31 A reinforced concrete stair 900 mm wide is centrally supported on a beam. Produce a sketch to show the arrangement of reinforcement in both the beam and step.

32 a) What are the uses of handrails on stairs?
b) Produce a neat sketch to show the method of fixing a handrail to the stair.

33 Describe two methods of making the surfaces of steps and landing non-slippery.

34 What is the purpose of a damp proof course in a building?

35 Describe three types of damp proofing materials and indicate the most appropriate positions where each can be used.

36 Sketch a section through a hollow ground floor and show the necessary construction details.

37 Produce a working procedure for damp proofing a basement.

38 Define the following types of damp proofing method:
a) tanking,
b) integral,
c) membrane,
d) surface treatment.

39 Describe the main difference between partition wall and party wall.

40 With the aid of sketches, show two ways of tying a partition wall to an external main wall.

41 State the purposes of the following building practices:
a) concrete wall head,
b) wall plate,
c) coping to parapets.

42 With the aid of neat sketches, illustrate the following:
a) a concrete wall head,
b) a pre-cast concrete saddled-back coping.

43 a) What is a gable?
b) Illustrate with sketches:
i) a verge gable,
ii) a parapet gable.

44 Produce sketches to show the method of building a small verge gable.

45 a) List four ways by which a block wall is likely to fail.
b) What are the general precautions to be taken against cracking of block walls?

46 List six different circumstances in which the use of retaining walls may become necessary.

47 In which major ways do masonry earth-retaining walls differ from reinforced concrete retaining walls?

48 List the functions of the following practices in masonry retaining walls:
a) stone coping,
b) weephole,
c) rendering to the back face,
d) attached piers,
e) granular filling at the back of the wall.

49 With the aid of sketches, illustrate the method of building attached piers to a retaining wall.

50 Illustrate with neat sketches, the method of reinforcing a sandcrete block retaining wall.

51 Define the following terms in relation to retaining wall construction:
a) angle of repose,
b) active pressure,
c) passive pressure.

52 a) State what is meant by the middle-third law.
b) Describe fully the significance of the middle-third law in the stability of retaining walls. Illustrate your answer with sketches.

53 A masonry retaining wall is 230 mm thick at the top, 690 mm thick at the bottom and 1·8 m high. Using the graphical method, determine the centre of gravity of the wall.

54 List five important points to observe when constructing a masonry retaining wall.

55 a) Describe, and illustrate with sketches, five possible ways in which masonry retaining walls may fail structurally.
b) Suggest precautions necessary to prevent the failures described in a).

10. Concreting

Concrete is a man-made stone made by mixing sand, broken stones and cement in pre-determined proportions, with sufficient water to enable the setting action of the cement content to take place. Concrete is presently one of the most popular materials for the structural parts of buildings and civil engineering works. When reinforced with steel, it has an almost limitless capacity for carrying loads.

Materials for concrete

Cement

Cement is the binding material in concrete. When water is added to cement a chemical action takes place resulting in the cement setting hard afterwards. Good concrete therefore depends largely on the quality of the cement being used. Cement in bags absorbs moisture if not stored well in a dry, air-tight place. Water absorption leads to loss of strength. A good cement which still retains its maximum strength is warm to the touch and contains no lumps of hardened powder. A lump of cement should not be beaten into powder and used for concrete because it has already lost its original strength. Cement should therefore be stored well, and the following method may be adopted.

Store all bagged cement in a watertight shed with a sound, dry floor. If the floor is not dry as is the case on most sites, construct a floor raised on boards standing on building blocks so as to raise the cement bags above the moist ground. In a large shed, cover the cement with plastic sheeting. Make sure that the roof of the shed does not leak.

On small jobs, cement may be stored in the open. Ensure that the cement is raised on timbers clear off the ground and covered with tarpaulin sheeting. Figure 145 illustrates these methods.

Nowadays, large contractors benefit from the use of cement silos for bulk storage of cement on site. The silos are portable and offer ease of handling. Silos are completely weatherproof, wastage due to broken bags is eliminated and pilfering which is often a major concern of builders, becomes impossible.

Fig. 145 Storing cement

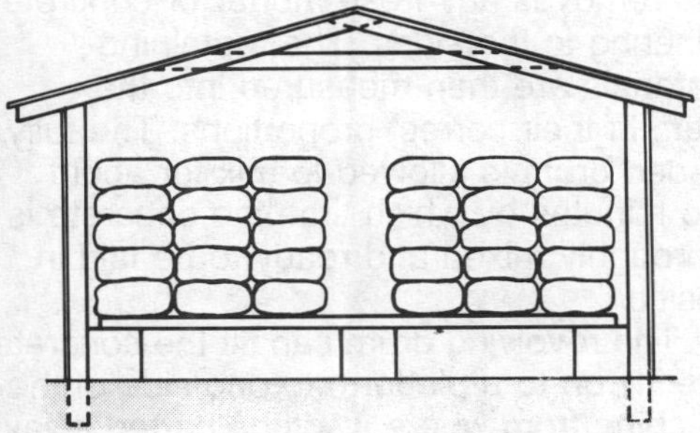

A typical arrangement of cement in 50 kg bags stored on site

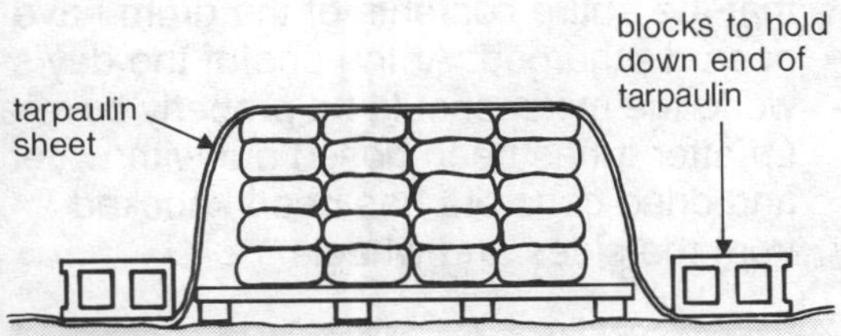

Method to be adopted when storing in the open

Fig. 146 Methods of storing cement

Sand and stone

These, known as fine and coarse aggregates respectively, should be clean when they arrive on the site for concrete work. They must contain neither animal nor vegetable matter nor lumps of clay, and must be well graded. When good quality natural sand can be found, it should be used as it is cheaper.

Reference should be made to the British Standards, B.S. 882: 1983 for concrete aggregates from natural sources. Nowadays quarries which produce stone dust out of granite may be found. This type of sand is expensive and is mostly used for high class work, such as bridges and road building. Stone dust does not have any of the problems of impurities associated with natural sand.

Aggregates must be tipped on to a hard, clean surface on the site. If there is no suitable surface, put down a layer of lean concrete and tip onto it. This extra cost will be more than made up for by the saving in aggregates later on in the job.

The amount of water in the aggregate directly affects the quality of the concrete. A wet mix of concrete produces a weak product so the water contained in the sand must be watched carefully. Dry sand is not easy to obtain for construction purposes during wet weather. Allow stocks of sand to stand at least twelve hours before using. This will allow the water to drain to the base and some moisture to evaporate into the air before the sand is used. Avoid taking sand from the bottom of the stock as this part will gradually become saturated.

Grading of aggregates

The word grading is used to describe the extent to which various group sizes of particles are present in a load of aggregate. Poorly graded aggregates contain particles of about the same size, while well graded aggregates contain particles ranging from the smallest size to the largest size. The definition of which size should be considered the largest is subjective to other things like whether the concrete being made is mass or reinforced concrete. It will suffice here to say that for mass concrete, the particle size may be 38 mm or more, but no particle should be greater in diameter than one quarter the thickness of the finished concrete.

The purpose of grading concrete aggregates is to produce a concrete that is readily placed and compacted, and to obtain concrete that is reasonably free of voids. There is no one ideal grading of aggregate for all concrete work. Grading of aggregate for a particular job is determined in advance by engineers having taken into account individual circumstances before a load is delivered to the site.

Bulking of sand

Bulking is the percentage increase in volume of sand when moist. Sand with a moisture content of 10% of dry weight will increase in volume up to 30%. Fine sand bulks more than coarse sand. It has been established, however, that a saturated quantity of sand has about the same volume as dry sand.

To avoid too much water in the concrete, the percentage bulk of moist sand must be ascertained and an allowance made for the increase in volume. This is done by measuring out a quantity of moist sand into a glass jar, noting the volume, and then saturating it with water and noting the decrease in volume.

Water/cement ratio

Water is an essential ingredient of concrete. It causes the chemical action which results in the setting and hardening of the cement and also enables the concrete to be sufficiently plastic for easy placing and compacting. While too much water in concrete makes the concrete weak and causes it to deteriorate fast, too little makes it difficult to place. The right proportion of water to cement must therefore be established. This is known as the water/cement ratio. The water/cement ratio is expressed by weight and for most concrete the limits of this ratio lie between 0·4m and 0·65m. Outside these limits, the concrete will either be too stiff or too wet.

Mixing concrete

Mixing concrete follows the same steps as mixing mortar. For small quantities concrete is mixed by hand and large quantities are mixed by machine. A gauge box (Fig. 48) is used for measuring the quantity of material when batching by volume. Weight batching is, however, a more accurate method. Many machines incorporate a weighing hopper for this purpose.

Hand mixing

Place the measured quantities of fine and coarse aggregates on a platform. (On site, the platform is usually a layer of lean concrete worked to a flat surface.) Each measured quantity is levelled off until the required overall quantity of aggregate is obtained.

Next, measure the cement, place on top of the heap of aggregates and spread evenly with the mixing shovel. The dry materials are mixed together with shovels, turning the mixture from side to side on the platform three times. When the heap shows an even colour throughout the mix is ready for the addition of water.

Water is added to a chasen formed in the middle of the heap preferably in sprays from a garden watering-can or a hose pipe. Water added from buckets on to the heap tends to wash away the cement in the mix before the mixer has time to mix the concrete. The mixture is further turned with shovels until it reaches a plastic state without being too wet. A well mixed concrete should be capable of standing in a heap. Concrete that flows and drains with water is unsuitable for most purposes. The concrete is now ready to be placed into position. See below right for slump test.

Machine mixing

Many small-sized contractors and private house builders are denied the advantages of machine mixing of concrete because of the high cost of purchasing or hiring mixers. Machine mixing is faster, saves materials and a better mixture is turned out.

The revolving drum type of machine is used for concrete mixing, the capacity of the machine being chosen to meet the required quantity of concrete on the job and the speed at which each batch can be laid.

There is no advantage to be gained by mixing the materials dry first so the required amount of water is first added into the drum. This moistens the drum and removes any fresh mortar or concrete adhering to the sides. The remaining materials are then measured into the drum in their correct proportions. The fully loaded drum is allowed to mix for about two minutes by which time the concrete is thoroughly mixed and ready to be laid in position.

The revolving drum can tilt the concrete directly on to a platform to one side of the machine from where it will be carted away in barrows or headpans, or directly into a waiting wheelbarrow or dumper. Check that the entire contents of the drum have been discharged. At the end of the day's work, the mixer should be properly covered up after it has been hosed out with water and dried concrete has been knocked from the sides and wheels.

Slump test and compacting factor test

In order to obtain an even strength in all parts of a concrete structure the amount of water in each batch must remain the same throughout the casting process. A batch of wetter concrete will result in a pocket of weaker concrete in the finished structure. A batch that is too dry will also result in a patch of conrete that cannot be easily placed and consolidated.

Excess water in concrete may also be caused by water already contained in the aggregate. Everything should be done to maintain consistency in the concrete. By visual observation, it is easy to tell if a batch of concrete is wetter than other previous batches. This visual test needs

Suggested proportions of cement, sand and stone	Types of work for which suited
1:1½:3	High strength watertight concrete
1:2:4	General use, e.g. walls, reinforced works etc.
1:3:6	Mass concrete: solid ground floors, foundations

Table 9 Suggestions for concrete mixes

to be checked by more reliable methods to ensure greater accuracy. Such methods include the slump test and the compacting factor test.

Slump test

The slump test is carried out to measure the distance that concrete compacted into a cone will slump down when the cone is lifted from it. The slump will be the same for the batches if the water content is consistent. Figure 147 illustrates the procedure to be followed when doing the slump test.

Compacting factor test

Workability is important in concreting work. Workability describes the ease with which a freshly mixed concrete can be laid, and to a large extent it is controlled by the amount of water in the concrete.

The compacting factor test measures the workability of concrete more truly than the slump test. It compares the weight of partially compacted concrete with a sample from the same batch that is fully compacted. It is intended to produce a figure as a measure of workability for various concrete mixes having the maximum aggregate size of 38 mm.

The apparatus used for the test is shown in Fig. 148. There are two hoppers, one below the other, with hinged flap bottoms. Below the hoppers is a cylinder. The test consists of two stages.

To perform the test, weigh the bottom cylinder empty, set in position and cover. Then mix the concrete thoroughly and gently place it in the top hopper to fill approximately without consolidation. Then release the top hopper flap and the concrete falls into the second hopper. Uncover the cylinder which has already been made ready to receive the concrete. Release the flap of the second hopper and the concrete falls and fills the cylinder. Some concrete is bound to spill over. Trim the concrete off the cylinder and, with a trowel or a similar tool inserted under, remove the cylinder and weigh it. Record the net weight. This is the first stage.

The second stage is to empty the cylinder, fill it with concrete with about

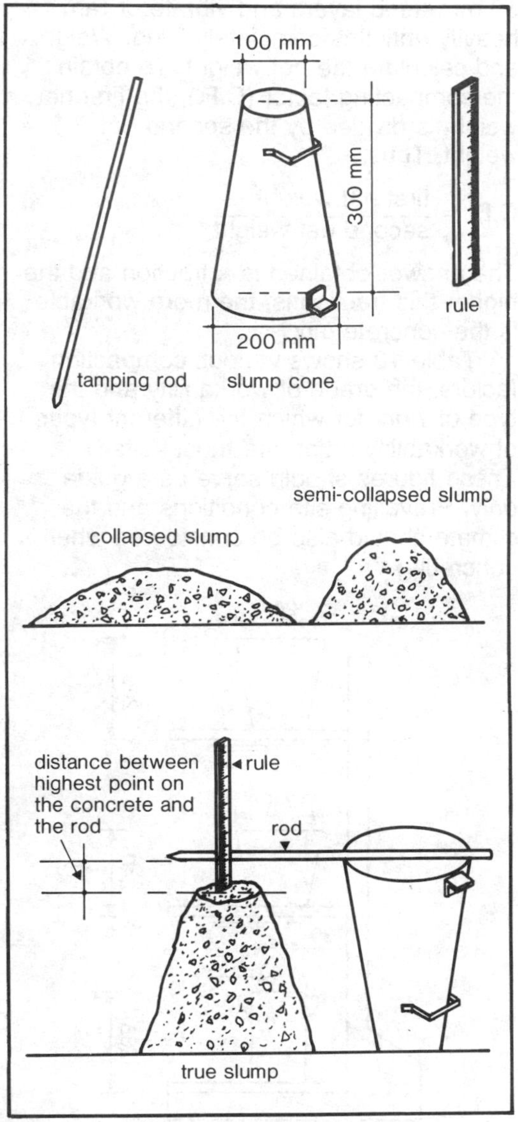

Fig. 147 The slump test

50 mm deep layers and vibrate or ram heavily until it is completely filled. Weigh, and calculate the net weight. To obtain the compacting factor (C.F.), the first net weight is divided by the second net weight. Thus,

$$\text{C.F.} = \frac{\text{first net weight}}{\text{second net weight}}$$

The answer obtained is a fraction and the higher this fraction is, the more workable is the concrete mix.

Table 10 shows various compaction factors, the grade of workability and the type of work for which the different types of workability rating are most suited. These figures should serve as a guide only. Prevailing site conditions and the climate should also be considered when concreting.

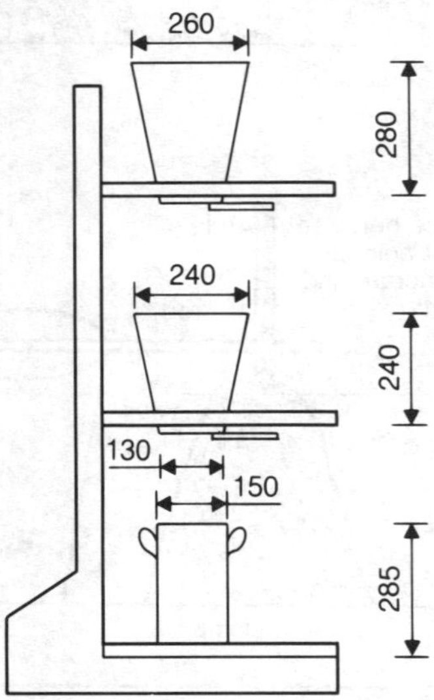

Fig. 148 Compacting factor test apparatus

Compacting factor	Workability	Type of work
0·95	Very high	Concrete work with complex reinforcement
0·90	Medium	Simple concrete works which do not need reinforcements
0·85	Low	Vibrated reinforced concrete in large masses
0·75	Very low	Vibrated large mass concrete

Table 10 Compaction factor, grade of workability and most suitable type of work

Transporting concrete

The system whereby the concrete is transported from the mixer to the point of placing depends on the type of job being done, the nature of the site, the equipment available, the distance to be covered below or above ground, and the cost of labour at the time. Concrete can be transported in the following ways.

Head pans

Concrete may be transported from the mixer to the point of placing by individual labourers carrying it in head pans or buckets. The number of people involved on each job depends on the size of the job, as this method is used for both small and medium sized jobs. This type of labour is most often used for the foundations of housing and the casting of floors up to first floor and roof. Hot weather and strong warm winds dry out concrete very quickly and this must be borne in mind when transporting concrete in this way. If concrete is not transported quickly after mixing, it may stiffen and become difficult to place. (See Fig. 149.)

Mechanical plants

These include powered barrows or dumpers, lorries, and cranes using skips. These are fairly fast means of transporting concrete. When powered barrows or dumpers are used, the route taken to deliver the concrete must be fairly even and be free of stones and broken blocks or other wastes lying about on the site. This is to prevent the concrete being segregated through vibration along the way. Lorries transport concrete over long distances and mixing and batching may be located outside the site. Concrete delivered this way is known as *ready mix concrete*, which is beyond the scope of our discussion at this level.

During very hot weather the moisture in the concrete can evaporate quickly. If the concrete is to remain out of doors for long it should be covered up.

Cranes where easily available are a convenient way of transporting concrete The cost precludes their use in concrete works, except on large contracts.

Wheelbarrows

Hand-pushed wheelbarrows are frequently used to transport concrete to the point of

Fig. 149 Transporting concrete in head pans

Fig. 150 Using a crane to transport concrete

placing, or to a hoist, from where the concrete and wheelbarrow are hoisted to the point of placing. On swampy or water-logged sites, a barrow-way may have to be constructed so that the wheelbarrow is pushed along a dry surface. This obviously slows things down. The barrow-way is usually made of planks placed end to end and firmly wedged in position. Again, if the wheelbarrow is to travel over rough ground, precautions must be taken against segregation of the concrete.

Placing and compacting concrete

Mixed concrete should be placed in its final position within half to one hour of mixing. This is because the initial setting of ordinary Portland cement takes place within this period of time. After this time, the concrete has lost some strength which can never be regained, and the load-bearing capability of the mature concrete is adversely affected.

Concreting in hot dry weather poses particular problems with regard to placing. Moisture evaporates too rapidly and this can result in a weak finished product unless due precautions are taken. Most architects specify that placing of concrete should be done soon after mixing and not later than half an hour after mixing. This specification is ideal and should be adhered to if best results are to be expected from the finished product.

When concrete is being laid in foundation trenches, in addition to the fast evaporation of moisture from the concrete, the dry trench absorbs moisture from the concrete. To limit this absorption, the trench should be watered down before the concrete is placed into it.

Concrete should not be dropped from a great height, as this tends to segregate the aggregates – heavy particles falling first. A height of 1 m is usually specified as the maximum from which wet concrete should be allowed to drop freely.

Soon after placing in the right position, concrete should be compacted or consolidated. The purpose of compacting is to make the concrete as dense as possible by eliminating voids within it. If a mix is too dry and stiff, proper compaction is not easily achieved. Avoid using too much force in compacting because not only may the reinforcements in reinforced concrete be dislodged from their positions and formwork forced out of alignment, but the stones may be broken into finer particles.

Consolidation may be achieved either manually by using a spade, a piece of iron bar or timber, or mechanically by employing a pneumatically operated poker vibrator which is inserted into the concrete while vibrating. A practice sometimes found on the site is that of putting the vibrating head of the poker vibrator against the formwork. This is bad practice because it damages the formwork, preventing it from being used as many times as possible. There is also the likelihood of setting the formwork lose with a devastating result. The exception to this is when it is not possible to insert a poker vibrator in the formwork for reason of it being too narrow or where there is heavy reinforcement. In such instances an electrical hammer with a flat vibrating end is used. The formwork must be strong enough to withstand the vibratory effect of the hammer.

All formwork must be plumbed, levelled and adequately supported, unless designed to be otherwise. Correct alignment of formwork is as important as providing adequate rigidity, especially where the concrete is being cast to receive prefabricated structures. Correct alignment is essential in order to avoid costly future corrective work. Wall and column forms are better aligned using the centre lines.

Curing concrete

This is a very important operation if the concrete is to attain its optimum strength. Curing means keeping the newly laid concrete under uniform conditions of temperature and moisture during the hydration of the cement compounds. Curing makes concrete stronger, more durable, denser and more resistant to abrasion. If concrete is allowed to dry out too quickly, the surface will be weak and more likely to dust off, cracks may develop and rain drops can eat into it. Direct sunshine and wind should be kept off newly placed wet concrete. This can be achieved by adopting one of the methods of curing discussed below.

Methods of curing

Plastic sheeting
This material prevents the moisture already contained in the concrete from escaping into the air. Cover the concrete shortly after placing, taking necessary care so that the sheeting is not blown away by wind. This is a very effective curing material.

Hessian or matting
After the concrete has been placed into position, these materials are used to cover it. The hessian or mats are wetted before placing, the aim being to keep water in close contact with the concrete until normal setting takes place.

Curing compounds

There now exist membrane curing compounds marketed under various names which offer specialised ways of keeping moisture inside the concrete until normal setting is achieved. They are usually applied 15 to 30 minutes after finishing the concrete. They are widely used in road works.

Formwork offers very good protection to the concrete from sun and wind, and is therefore a good curing aid although often the form is removed for re-use only hours after placing the concrete. Plastic sheeting, damp matting or hessian are then wrapped round columns and beams.

Sand layer
Sometimes a thin layer of sand is used to cover the surface of the concrete. The drawback in this method is that one must necessarily wait until the concrete has achieved some considerable setting before the sand is spread over it, to avoid

causing bruises to the surface, or embedding the surface with sand.

Water spray
Where there is an ample supply of water, a hose pipe could be used to spray the concrete with water, keeping it moist for days. While this is an excellent method of curing, water shortages in many areas usually preclude its use.

Period of curing

The period during which concrete must be cured is specified by the engineer. Generally the curing time is the period for which the concrete must remain covered or moistened and must not be taken to mean the period after which the concrete can be put to use, or the time before the concrete will have all supports removed. These periods should also be specified.

Types of concrete

Broadly, concrete can be divided into two types: mass concrete and reinforced concrete (r.c.).

Mass concrete

This type contains no reinforcement. Most foundations on firm soils which will not carry heavy loads are of this type. Structures which have to be heavy in order to be stable are also made of mass concrete, e.g. concrete gravity dams and some retaining walls with only compressive stresses in them.

Reinforced concrete

Concrete is strong in compression but weak in tension. Because concrete is used to withstand various stresses, there is a need to introduce steel into its mass to fulfil such functions. Steel combines very well with concrete to act against forces. The steel resists tensile forces while the concrete resists compressive forces. The reinforcement in concrete may be simple bars, welded mesh fabric or a series of rods bent and tied to a given schedule with wire stirrups. The purpose of bending and the use of stirrups is to enable the concrete to resist shear stresses and also to keep the bars in their correct positions until the concrete is laid.

Types of reinforcement

Plain round bars

These are of mild steel. The bars are made in various diameters ranging between 4·8 mm and 50·8 mm. They are very commonly used and their manufacture is controlled by B.S. 4461. (See Fig. 151.)

High bond and twisted bars

These are manufactured to conform to B.S. 8110: 1985. The twisted bars are made from square section bars cold twisted about their longitudinal axis to a predetermined pitch. The cold twisting of the bars hardens the steel, and this increases the working stress of these bars by at least 50% over that of ordinary mild steel.

High bond bars have small projecting transverse ribs usually in a herringbone pattern, and a longitudinal rib or groove.

Advantages associated with the use of these types of reinforcing bars are that they offer superior bonding working stresses and reduce the amount of steel required. Close supervision is, however, needed to produce a superior quality concrete that will match the high strength of the steel. They are also more difficult to work and raise the cost of labour and materials over that of ordinary round mild steel bars. (See Fig. 152.)

Mesh fabric

This consists of series of longitudinal main wires held at fixed distances apart by transverse wires welded at their intersections. It is obtainable as a square or a rectangular mesh, and in flat sheets or in rolls. (See Fig. 153.)

Expanded metal fabric

This is variously called 'EXMET' and 'XPM'. It is made from a steel sheet which has been cut through the surface and expanded to form a diamond-shaped mesh (Fig. 126).

Anchorage and bond of reinforcement

There is little to be gained from reinforcing concrete if there is not enough bond between it and the concrete. A badly

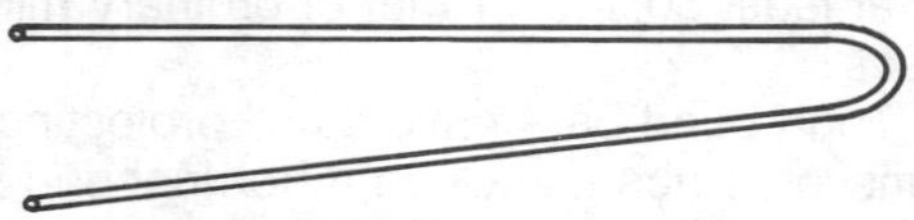

Fig. 151 Round mild steel bar

Fig. 152 Twisted bar

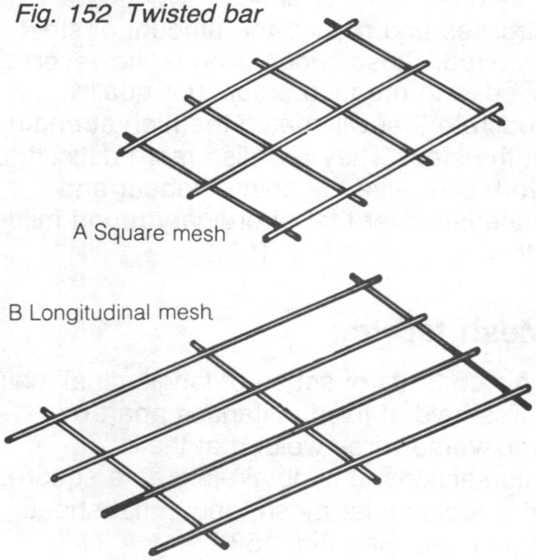

Fig. 153 Welded mesh fabric

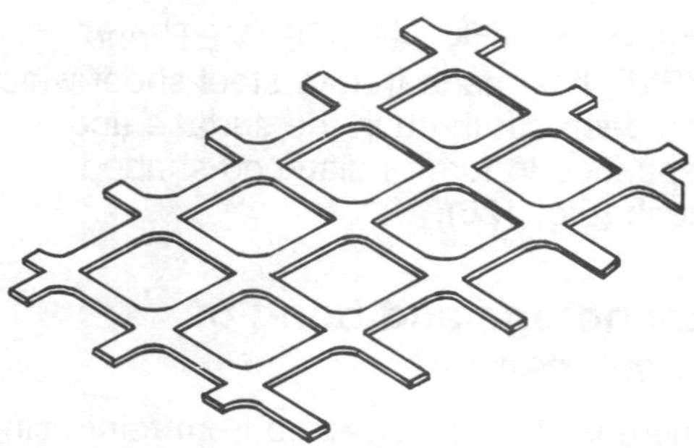

Fig. 154 Expanded metal

Fig. 155 A heavily reinforced upper concrete floor showing a good example of high bond reinforcing bars being used in beams and columns

rusted reinforcement should not be used for reinforced concrete works. The ends of reinforcing bars must be adequately anchored in the concrete so that there is no possibility of slipping in case of failure of the structure. There are two recognised ways of bending the ends of reinforcing bars: a 90° bend upwards or a hook bend, referred to as an 'L hook' and a 'U-hook' in B.S. 8110.

Concrete cover

Mild steel reinforcement corrodes under moist conditions. Once the steel is inserted into the concrete, it should not come in contact with air and water. Sufficient cover should therefore be provided for all reinforcements including links, stirrups, etc. B.S. 8110: 1985 states the following covers.

1 For each end of a reinforcing bar, not less than 25 mm, nor less than twice the diameter of such bar.

2 For a longitudinal bar in a column, not less than 40 mm, nor less than the diameter of such bar. In the case of columns with a minimum dimension of 200 mm or under, whose bars do not exceed 12 mm in diameter, 25 mm cover may be used.

3 For a longitudinal reinforcing bar in a beam, not less than 25 mm nor less than the diameter of such bar.

4 For tensile, compressive, shear or other reinforcement in a slab, not less than 15 mm nor less than the diameter of such reinforcement.

5 For any other reinforcement the cover not to be less than 15 mm nor less than the diameter of such reinforcement.

For all external work, for work against earth faces and also for internal work where there are particularly corrosive conditions, the cover of the concrete should not be less than 38 mm for all steel, including stirrups, links, etc. except where the face of the concrete is adequately protected by a suitable cladding or by a protective coating, which may indeed be advisable where the corrosive conditions are unduly severe.

Ways of casting concrete

There are two ways by which a concrete product or structure can be made, i.e. casting the concrete in the position of final rest or casting it elsewhere and later bringing it to the position where it will be fixed on the job. The former is called cast-in-situ concrete and the latter is called pre-cast concrete. The rules and procedures governing the use of reinforcement should be adhered to irrespective of whether the concrete is cast-in-situ or pre-cast.

The decision whether or not to adopt a particular way is influenced by:

1 the weight of the concrete structure,

2 the distance between the casting shed and the site where the product will finally be fixed,

3 the type of structure being constructed,

4 whether or not the site has enough room for the storage of the materials.

There are advantages and disadvantages of both ways.

Advantages of pre-casting

1 Construction work is not interrupted by bad weather.

2 There is speed in construction.

3 There is less risk of reinforcement being displaced.

4 Better finish is possible.

Disadvantages of pre-casting

1 Transportation and handling costs are incurred.

2 Member may be damaged through incorrect or repeated handling.

3 Jointing may be a problem.

Pre-stressed concrete

This is a type of reinforced concrete whereby high tensile reinforcements are stretched before placing the concrete so that the stretched reinforcements trying to return to their original lengths cause a compressive pre-stress in the concrete. A concrete member treated this way has greater load-bearing potential. The stretching of reinforcement before it is cast into concrete is known as pre-tensioning and stretching the reinforcement after the concrete has been cast is known as post-tensioning.

Formwork and moulds

The construction of timber formwork is strictly the carpenter's job but it is essential for the blocklayer to be acquainted with the basic principles of construction, so as to be able to appreciate its use fully.

Moreover, he should be able to construct simple forms and moulds himself. Nowadays there are, in addition to wood forms, proprietary systems of formwork which only need assembling on the job.

Formwork is a general term used to describe the preparation and erection of moulds for concrete which is cast-in-situ or indeed the moulds themselves. Moulds are boxes or castings into which the concrete is poured to form pre-cast units.

All formwork must be:

1 sufficiently strong and rigid to support the dead load of the concrete,
2 designed to allow for the temporary live load of workmen, wheelbarrows and tamping operations,
3 easy to strip after concreting,
4 constructed within reasonable cost limits. In tropical climates, a problem sometimes experienced is the rapid absorption of moisture from the concrete by timber members of formwork. As emphasised earlier, rapid withdrawal of moisture from wet concrete is harmful so partially seasoned timbers are preferred. Seasoned timber will require wetting before concrete is placed, so that moisture will not be absorbed too fast from the wet concrete.

Types of formwork

Timber is the most commonly used material in the construction of formwork. Timber formwork is usually made from softwoods free from excessive knots and other defects. It is easily cut to shape, easy to fix and dismantle. (See Figs. 100 and 101)

Plywood is used where a large area of exposed concrete is to be constructed, e.g. when constructing walls. The advantages associated with the use of plywood in formwork inclue the elimination of joint marks and the fact that, if carefully fixed on site, the finished concrete does not require further treatment.

Steel can also be used. Most steel formwork is patented and part of proprietary shuttering systems with special panels, fixings and supporting devices. The advantages of this type of formwork include: speed of erection; fewer men required to assemble it on site; better surface finish of concrete; can be used several times, resulting in substantial savings compared with timber formwork.

Steel is initially more costly than timber formwork, but because it can be re-used, it is eventually more economical. Only larger firms can afford steel forms because of the initial outlay.

Re-use of formwork

The cost of formwork in concrete work is considerable. It is therefore important to consider at the design stage how often the formwork can be used. As formwork takes a long time to make, it would be uneconomical to make a formwork that could only be used for one job. Concrete members should therefore be made as similar as possible to reduce the types of forms required and to ensure maximum re-use. Beams, columns, floor slab thicknesses and their spacings should be made identical on all floors.

Finishes applied to concrete

Exposed concrete surfaces are often either painted or left as cast. Painting requires periodic renewal especially in dusty conditions such as occur in tropical countries. It is, however, possible to give a concrete surface a pleasant appearance without painting. This can be achieved by one of the following means.

Exposed aggregate finish

This is achieved by removing the fine aggregate chemically or mechanically from the concrete surface to expose the larger aggregate. This produces a rugged natural appearance.

Applied finishes

These include plastering and painting, tyrolean and spatter dash. These finishes are applied in accordance with mixes and methods specified by the architect.

Board marked finish

A variety of attractive finishes can be obtained by using rough sawn timber for the sides of the concrete formwork. Another appearance can be obtained by chamfering the edges of the individual planks. This leaves continuous small projections on the surface of the concrete. One can also fix cleats to the internal face of the formwork to obtain the reverse feature. Alternatively corrugated metal sheets can be used. They come off easily and leave a beautiful finish on the concrete.

Board linings

Various materials, such as ribbed hardboard, coconut mat, plastic sheeting, hessian, specially shaped plastic panels, etc. can be attached to the inner face of the formwork to produce different attractive surface finishes.

Revision questions

1 a) What is the difference between a mass concrete and a reinforced concrete?
b) In what circumstances will mass concrete be recommended in preference to reinforced concrete?

2 What are the necessary checks that have to be carried out to ensure that the following materials will give the best result when used for concrete:
a) cement,
b) sand,
c) stone,
d) water?

3 Describe, using sketches where possible, the methods of storing cement bags on a building site.

4 Define the following terms:
a) grading,
b) bulking,
c) water/cement ratio,
d) segregation.

5 List three ways by which a mixed concrete can be transported on the site. What precautions are necessary when concreting in hot weather?

6 Describe the method of mixing concrete by hand.

7 Write a working procedure for laying a large concrete floor.

8 Define the following terms:
a) curing,
b) bottoming,
c) compaction.

9 Describe three different ways of curing newly laid concrete.

10 a) Illustrate with the aid of sketches, the following types of reinforcement:
i) round mild steel bar,
ii) mesh fabric,
iii) expanded metal.
b) For what situation is each of the reinforcements in a) best suited?

11 What is pre-stressing? What are the advantages that a pre-stressed concrete has over a traditionally reinforced concrete?

12 Describe two methods of surface finish for a concrete wall.

11. Floor construction

A floor is a horizontal structure which carries imposed and live loads in a building and divides a building into storeys. The most common materials used for the construction of floors that will meet the requirements of building regulations and local bye-laws are concrete and wood.

The construction of upper floors is outside the scope of this book. We are therefore limiting our discussion to the construction of ground floors.

Ground floors may be of hollow or solid construction. Solid ground floors are of two main types, the traditional rammed earth and concrete.

Rammed earth floors

The soil used for this type of floor is clay or a clayey soil, which must be capable of being rammed into a reasonably solid state. It is dense with a surface that holds together. The floor is used mostly in rural areas where house builders may not be able to afford cement for the construction of their houses. It is important to know about this type of floor construction because when something is done it ought to be done well. Moreover, quite often blocklayers are called upon to advise on the construction of rammed earth floors or even to lay one themselves.

In a rural community, the floor of a building is not used as much as that in an urban community. The reason, of course, being that most people work on their farms for most of the day. When they do stay at home, their leisure hours are spent out of doors at the community square or under the shade of trees in the grounds of the house. The earth floor therefore does not normally suffer rapid wear as would be the case if it were more constantly used. However, a poor choice of soil will still result in rapid deterioration of the floor.

The procedure for laying the floor is similar to the filling and ramming of hardcore beds for a concrete floor. The original soil within the floor area is spread evenly with a shovel to a fairly level surface and properly consolidated. Special attention must be paid to the intersection of the floor with the walls, as this is the area most prone to failure.

The floor is laid in two layers each 150 mm thick. The first 150 mm (the base) is made up of clayey soil. Where there is an abundance of ballast in the locality, it is used for the base. This is levelled and rammed as previously described. Water is sprinkled on if the material is too dry, to keep the material moist during the consolidation.

The top layer is then laid. It is made up of clayey soil in a moist state. The right amount of moisture is essential in order to avoid excessive drying shrinkage and cracking in the finished floor. The soil must be capable of caking under pressure in the palm. A clayey soil in this state should give the type of finish which meets an acceptable standard. To improve resistance to surface wear due to abrasion, a quantity of dry cement powder is evenly spread on the top and thoroughly worked in with a hand rake. The material is then consolidated using a wooden tamper, as shown in Fig. 156.

During this final consolidation the floor is worked to a fairly flat surface by cutting and filling. A smoother, which may be a length of aluminium sheet or similar material attached to a handle, is used to smooth the floor by beating. A fairly smooth finish can be obtained with this simple tool if the soil contains the correct amount of moisture. It must be neither too dry, because a dry soil proves very difficult to work to a smooth surface, nor too wet, because a wet one will tend to be resilient and lift during smoothing.

The floor is then 'cured' by covering with grass, banana leaves, palm leaves or similar material. Curing takes about eight days. If the building was roofed before the floor was laid, the curing materials will not be necessary, as the roof will

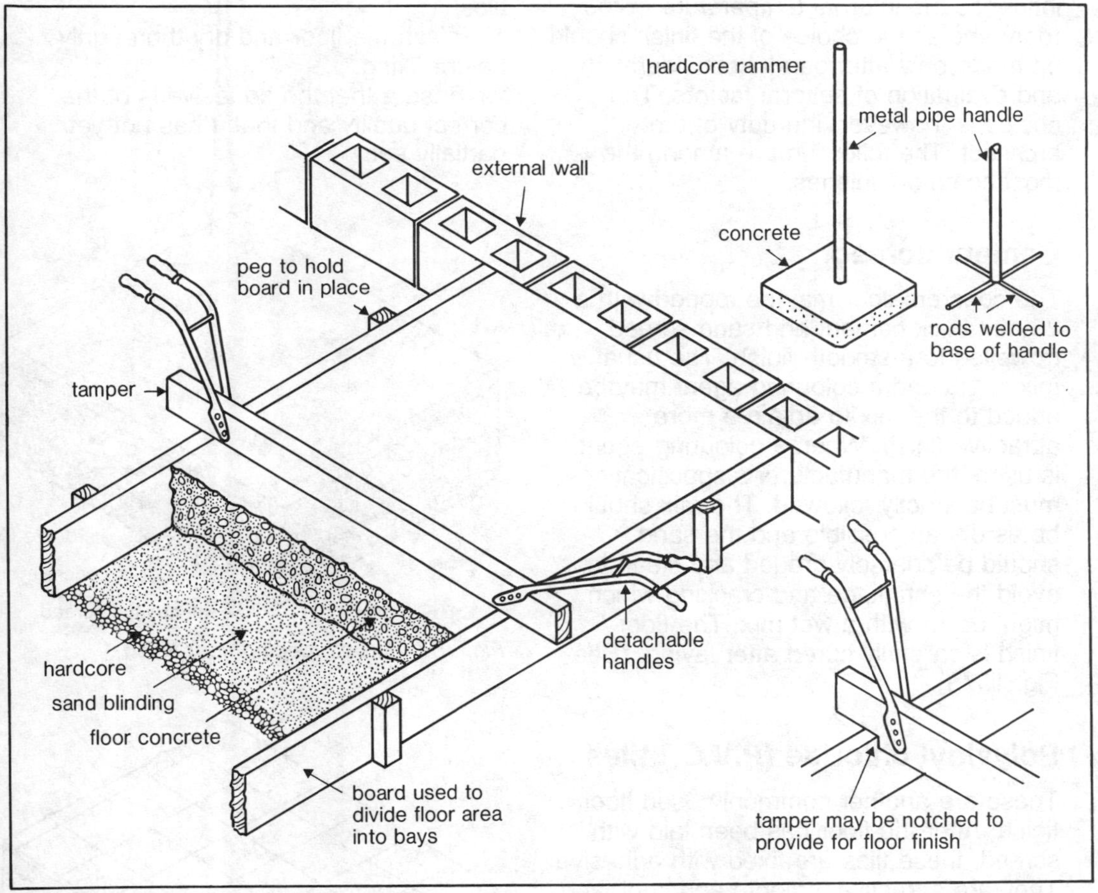

Fig. 156 Concrete and hardcore tampers

adequately protect the newly laid floor from direct sunshine until it is 'cured'.

Apart from cement, there are now many soil stabilisers available on the market, any of which can be used to improve the wearing qualities of the earth floor and to prevent it from cracking.

Concrete ground floors

Concrete ground floors are the most popular type of floors for single storey buildings in many tropical countries. They are durable, hard-wearing and cool to the touch, ensuring reasonable comfort for the users. They also provide a good barrier to the penetration of moisture into the building from the ground.

Ground floors may be of hollow or solid construction. Each type is described below.

Solid ground floors

This type is formed from mass concrete laid directly on the ground or on a filling of hardcore. The practice in some areas is to place a 100 mm thick layer of concrete at least 150 mm above the highest point of ground to cover the entire area of the building enclosed by the walls. This concrete layer is made to project about 50 mm beyond the external faces of the walls. It is claimed that this practice prevents dampness and termites from entering the building. While such a layer of concrete is capable of preventing dampness in the room in most areas, the second claim of prevention of termite entry into the building is doubtful, as the ground is not the only point where termites enter. Should it be known that termites are common in the area, more elaborate precautions should be taken to prevent termite damage. Some of these precautions are dealt with on p. 100.

For ordinary domestic purposes, the thickness of concrete mentioned above is sufficient, but for industrial and commercial buildings, the ground floor requires a greater thickness and might have to be reinforced.

Procedure for laying a ground concrete floor

When the walls have reached the damp proof course (D.P.C.) level, about 150 mm above the highest point of ground, the room spaces are levelled up to a depth of about 150 mm from the top of the walls. This may mean reducing the soil.

1 Fill the room spaces with high quality hardcore. This can either be laterite or ballast, broken blocks or stones.
2 Ram down the hardcore until it is level with the walls.
3 Lay D.P.C. over the floor area and walls if dampness from the ground is likely to be troublesome in future. This may be wallpaper, plastic sheeting or any other suitable D.P.C.
4 Fix formwork against the external sides of the walls to leave a depth above the wall equal to the thickness of the concrete.
5 Wet the rammed hardcore and walls thoroughly and treat the internal faces of the formwork with used motor oil.
6 Place mixed concrete and tamp down well.
7 Work concrete to a flat level surface with a straight edge.
8 Cure concrete as usual until it gains enough strength.

Finishes applied to solid ground floors

There are various finishes that can be applied to solid concrete ground floors. The type of finish given to the floor will influence the internal temperature of the room and so the choice of the finish should be made only after careful consideration and evaluation of several factors. The choice is, however, the duty of the architect. The following are among the most common finishes.

Cement screed

The concrete floor may be topped with a 25 mm thick cement and sand screed trowelled to a smooth finish. The usual mix is 1:3 and a colouring agent may be added to the mix to obtain a more attractive finish. When a colouring agent is used, the manufacturer's specifications must be strictly followed. The mix should be as dry as possible and the sand should be coarsely graded and clean to avoid the shrinkage and cracking which might occur with a wet mix. The floor finish is carefully cured after laying. (See Fig. 157.)

Polyvinyl chloride (P.V.C.) tiles

These are another commonly used floor finish. After the floor has been laid with screed, these tiles are fixed with adhesive. They are attractive, smooth and cool, and damage can be repaired very easily as they are made in small square size, usually 150 mm and 225 mm (See Fig. 158.)

Because of dust and the poor workmanship sadly common in construction work, this type of finish often fails after a few years of fixing, mostly through lifting. The following precautions are therefore necessary when fixing the tiles.

1 Wash the floor and dry thoroughly before fixing.
2 Ensure that the adhesive is of the correct quality and that it has not yet partially set.

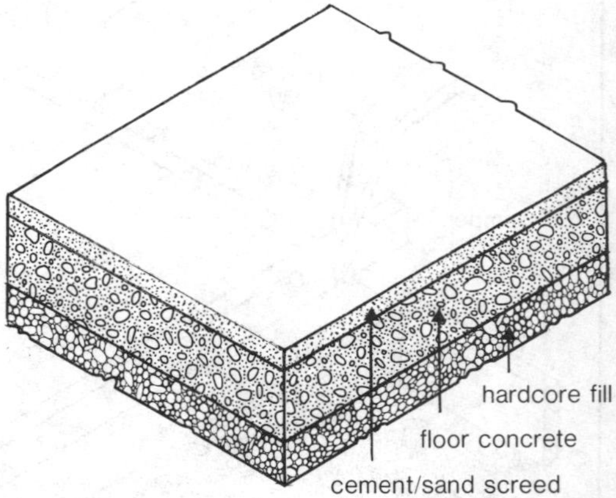

Fig. 157 Screed topping floor finish

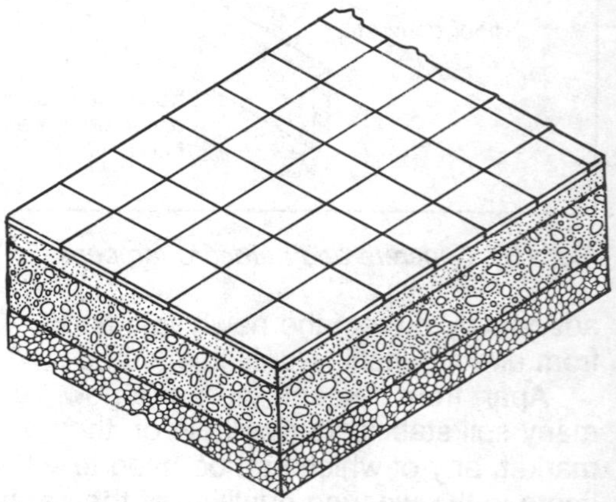

Fig. 158 P.V.C. tiled floor finish

3 Work out the number of tiles and place odd strips at the ends next to the walls.
4 Ensure that the floor is completely covered with the adhesive, leaving no patches.
5 The gap between the tiles should be as narrow as possible. Many finishes of this type fail because the gaps between the tiles are too wide and thus allow dust to accumulate in them and gain access under the tiles thereby lifting the edges.

Terrazzo

Terrazzo consists of a coloured cement binder or matrix and marble chips mixed to specified proportions. The finish is hardwearing, attractive and resistant to chemical attack. Terrazzo, sometimes referred to as 'granolithic' finish behaves in the same way as concrete. Ebonite strips are used to divie the terrazzo into bays to resist expansion and shrinkage.

The durability of all floor finishes depends upon the bond that exists between them and the subfloors. In laying terrazzo *in-situ* the floor is divided into bays of convenient sizes using the ebonite strips bedded in cement mortar. After setting, clean the subfloor thoroughly and brush the bay to be laid with cement grout. To minimise excessive differential movement of the finish and the subfloor it is better to lay the terrazzo while the concrete subfloor is still 'green'. Unfortunately, this is not often possible. The material is mixed and laid. It is ground and polished to a smooth surface after drying. The grinding exposes the marble which may contain various colours. Terrazzo may be supplied in tile form and laid by bedding in cement mortar. (See Figs. 159 and 160.)

Fig. 159 Grinding a newly-laid terrazzo floor

Woodblock and floor boards

The solid floor may have woodblock bedded directly on top with a bituminous adhesive or dove-tailed fillets bedded in the concrete and later finished with tongued and grooved boards as shown in Fig. 161.

These finishes are warmer than the previous ones discussed. This plus the cost of maintenance may preclude their use in ordinary housing schemes. A suitable damp proof layer must be provided under the ground solid floor in damp (e.g. marshy) areas.

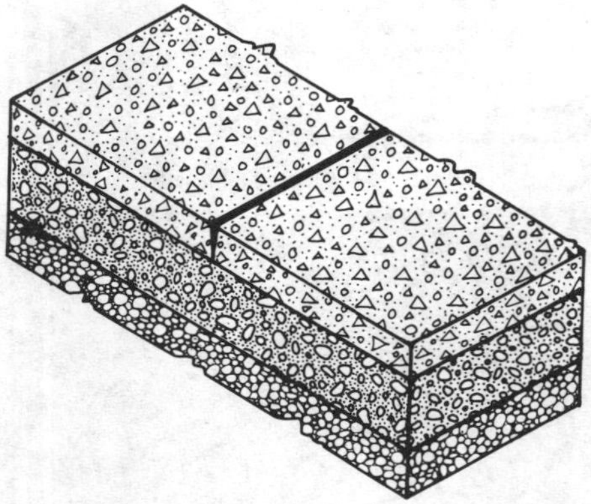

Fig. 160 Terrazzo floor finish

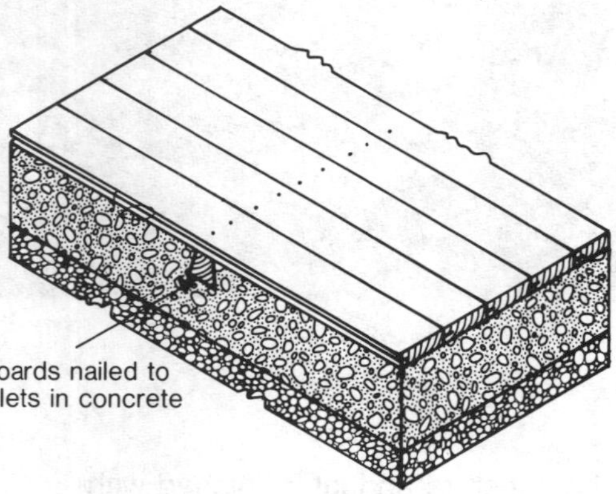

Fig. 161 Wood board finish

Hollow ground floor

A hollow ground floor is necessary where the ground is very wet and water prevention would be expensive. The floor consists of a series of narrow beams called joists, which are supported at regular intervals of 1·36 m by honey-combed sleeper walls, and covered with tongued and grooved boarding. The floor space below provides insulation against damp, but it is imperative that a D.P.C. be inserted below all timber works.

Ventilation

The pocket of air trapped within the space below the floor becomes stagnant after a while. This results in condensation and creates conditions where fungoid diseases can breed and attack the timber. Adequate ventilation must therefore be provided within the underside of the floor. This is achieved by building air bricks or blocks in the external walls, through which a constant air flow into the underside of the floor is maintained. To facilitate the free circulation of air within the floor the sleeper walls are honey-combed as shown in Fig. 124.

Site concrete

The site concrete is a weak concrete laid under the hollow floor to stop the growth of vegetation below the floor. It also serves as a foundation for the sleeper walls. (See Fig. 124.)

Termite protection

Termites are a menace to wood in most tropical areas, and all necessary precautions must be taken to protect wooden parts of a building from attack. The most reliable treatments are organic solvent preservatives, creosote (tar oil) etc. The wooden members are brushed over with the solution after being fixed in position. The framework of the timber floor is treated before the boarding is fixed.

Revision questions

1 Describe a method of constructing a rammed earth floor.
2 With the aid of suitable sketches, illustrate two types of hand tampers.
3 List four kinds of concrete ground floor finishes and write a working procedure for applying one of them.
4 How can termite attack on wooden members of a building be prevented?
5 What is the purpose of site concrete?

12. Masonry

Masonry is the art of shaping and laying blocks of stone to form walls. Masonry, which may be a separate craft elsewhere, forms an integral part of blocklaying and concreting in some countries. Stonework is, however, more costly than other materials, even in places where stones can be obtained easily. The extra cost arises from labour and the time taken to construct stone walls.

Even though there is an abundance of building stones in some areas, house owners often elect not to build with stones, even if they can afford the cost. This is in spite of the immense advantages that can be derived from the use of stones.

Advantages of stone walls

The advantages of stone walls over sandcrete block walls are as follows.

1 More durable under normal conditions.
2 Less likely to develop cracks due to expansion.
3 Greater load-bearing capacity per square metre of walling.
4 Do not require further external surface treatment, such as plastering or painting, therefore lower finishing costs.
5 More attractive natural surface finish.

Masonry tools

The mason uses the same laying, levelling and wall straightening tools as the blocklayer. These were described on pp. 13–18. We shall only deal with those tools that are of particular interest to the mason; such tools include implements for dressing and shaping stone.

Nowadays, there are items of machinery, both fixed and portable, which are used to dress and shape stone, leaving the mason only to trim and work the stone to finer details. However, as not all builders will be rich enough to purchase either already cut-to-shape stones or the machinery with which to shape and dress the stones, it is still essential to describe to the younger craftsman the tools he can purchase for his tool box.

To a certain extent, the particular type of tool used is determined by the hardness of the stone to be worked. The following list is not comprehensive but the tools described are necessary in the mason's tool box.

Stone punch
This has a small chisel-shaped cutting edge, and is used together with a hammer for rough dressing of the stone. (See Fig. 162.)

Pitcher
The pitcher has a wide and thick cutting edge, and is used for reducing the stone to a fair face. (See Fig. 163.)

Mason's hammer
This is similar to the normal block or brick hammer except that the steel head is much longer. The hammer may be used to chip off superfluous stone in rough dressing (See Fig. 164.)

Boaster and tooler
These tools are used in conjunction with the mason's *beechwood mallet*. They have wide striking ends which prevent the mallet from damage. The tools are suitable for soft stone work. (See Fig. 165.)

Claw
The claw has a series of teeth which prevent the surface of the stone from being holed during dressing. It is used mainly for drafting and dressing. (See Fig. 166.)

Steel straight-edge
This is used for aligning blocks of stone during the building process. The steel straight-edge is not a substitute for the spirit level or plumb rule, rather, it should be used in conjunction with either of the two. (See Fig. 167.)

Fig. 164 Mason's hammer

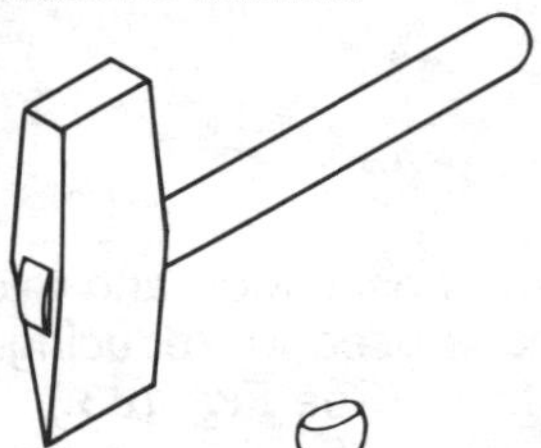

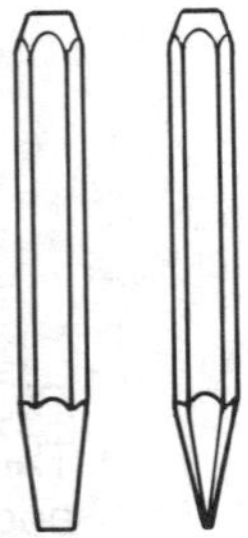

Fig. 162 Punch

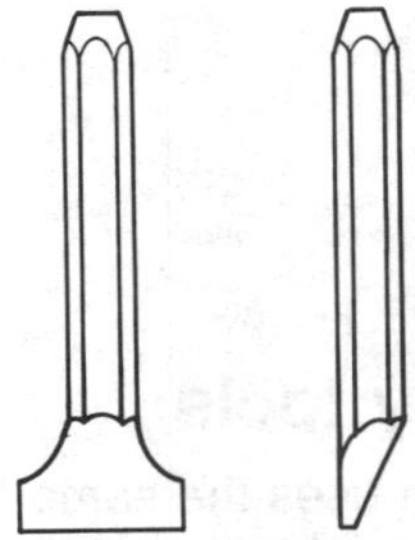

Fig. 163 Pitcher

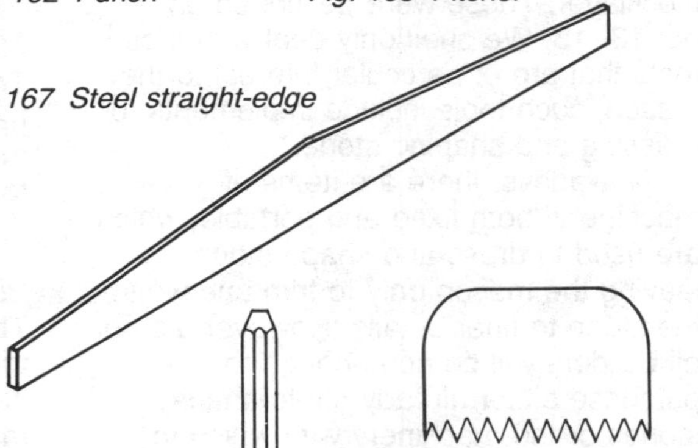

Fig. 167 Steel straight-edge

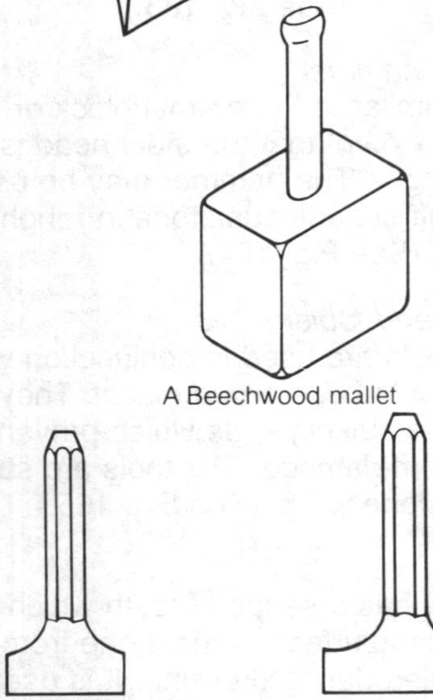

Fig. 165 Mallet, boaster and tooler

Fig. 166 Claw

Fig. 169 Drag

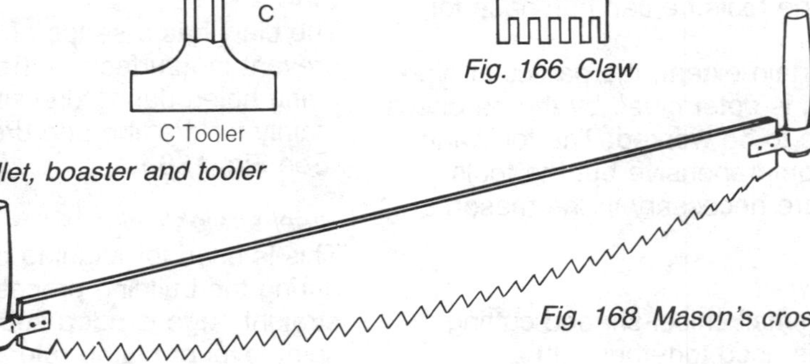

Fig. 168 Mason's cross-cut saw

Mason's cross-cut saw
This is normally between 2 m and 2·13 m in length and 250 mm wide at the centre. The double handed saw is used for cutting large blocks into sizes and requires two people for use. (See Fig. 168.)

Scratcher, comb and drag
These are abrasive surface finishing tools used on soft stones. They have saw-like teeth on the working edge. These tools can easily be made from sheets of spring metal to suit a particular use. (See Fig. 169.)

Bonds and bonding for stonework

Like brick and blockwork, a stone walling is known by the shape of its face bond. The most common bonds in stone walling are as follows.

Ashlar

This consists of stones worked to perfect shapes, the blocks having arrises like sandcrete blocks. They are laid in regular courses of 300 mm or more deep. Ashlar work is costly and it is usually constructed with rubble work or block backing. The mansonry is tied to the backing with metal cramps or 'block-bonded' with the backing walling by interlocking. Modern ashlar work uses metal cramps because this is economical. Ashlar work is illustrated in Fig. 170.

Flint

This type consists of small irregular sized split stones. It is usual to bond flint with brick, slate or bigger stone courses. The quions of flint walls are built with square stones or blocks for greater rigidity. (See Fig. 171.)

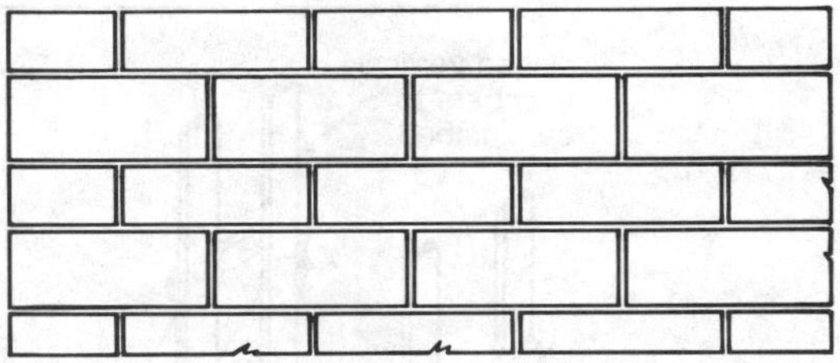

Fig. 170 Ashlar work

Fig. 171 Flints

Rubble

Rubble is used as a term to describe any stonework which is not highly finished. The chief types of rubble are as follows.

Squared rubble
This consists of roughly dressed stones with square edges. The stones may be square in form or rectangular. The courses are brought to a level bed joint at reasonable intervals (Fig. 172B).

Random rubble
This consists of irregular shaped stones of various sizes. The stones are laid randomly, i.e. big stones and small ones are laid next to each other with no attention paid to their levels as shown in Fig. 172C. Random rubble may be laid in courses and, whether or not stones are laid in courses, they must be brought to the same level at intervals.

Rag rubble
Rag rubble presents an attractive surface appearance with roughly hexagonal main stones and other irregular-shaped minor stones (see Fig. 172D). Reference should be made to pp. 105–106 for photographs of some of these stone walling types.

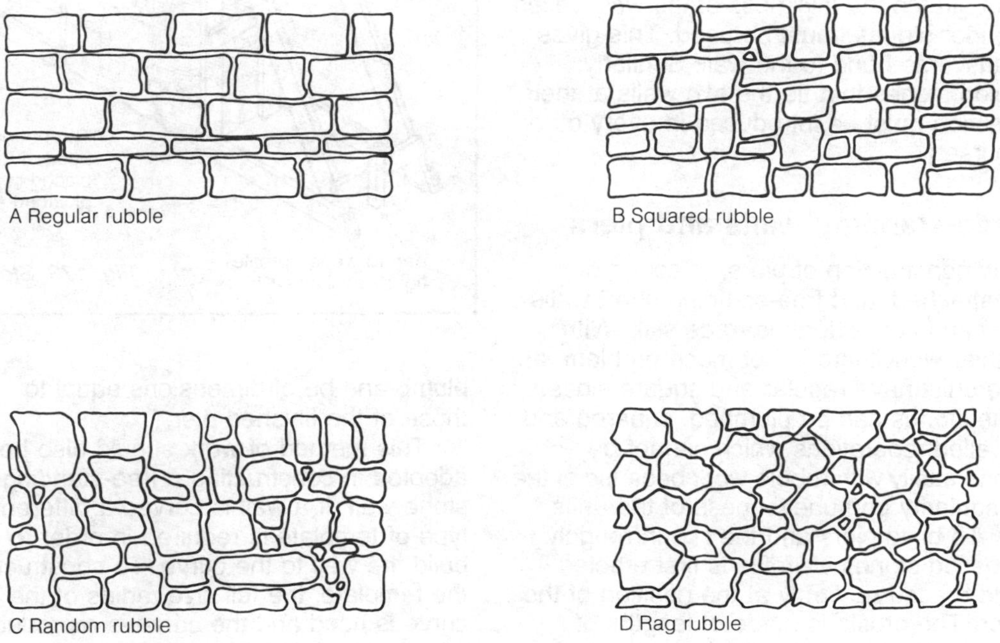

Fig. 172 Rubble

Walling techniques

Stone walling demands a great deal of judgement, forethought and patience in the selection and laying of the stones. The good mason can make a first class job out of the most unusual heap of broken stones. To him every edge of the stone has its artistic purpose, and he utilises it to advantage.

The larger stones are placed on the two sides of the wall, while smaller ones are used to fill up the middle. This filling of the middle with small stones is referred to as hearting. As often as possible stones that stretch the thickness of the wall, called bonder stones, must be used. This gives transverse bond to the wall. Similarly, quion stones that tie the two walls at their junction must be introduced in every other course.

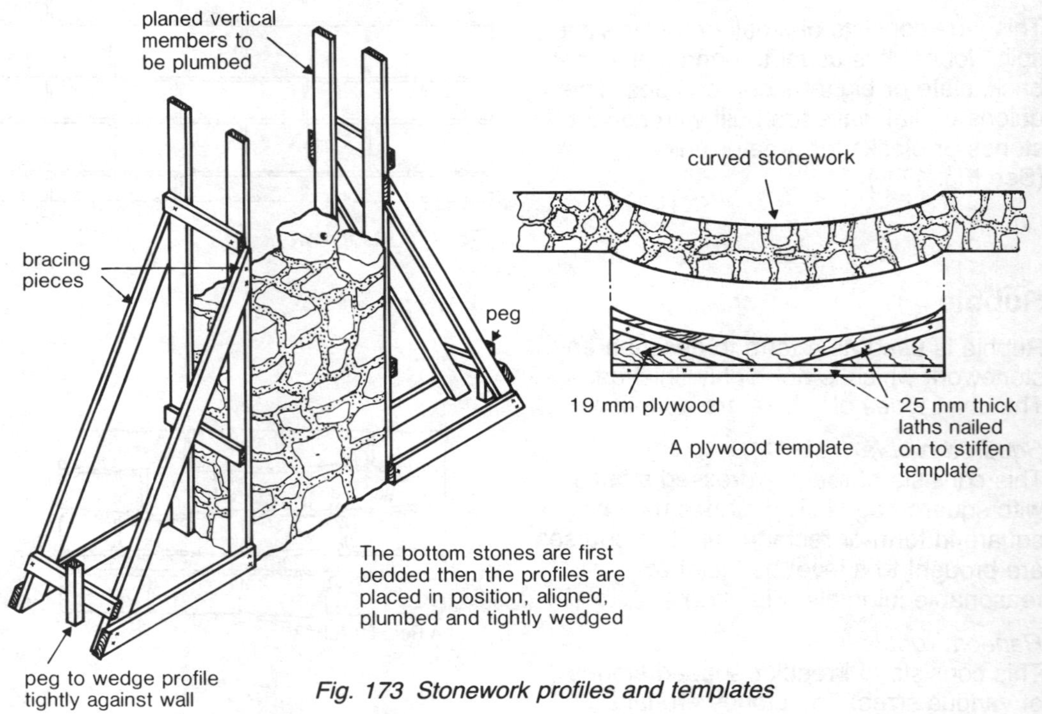

Fig. 173 Stonework profiles and templates

Free-standing walls and piers

The construction of piers, attached or unattached, and free-standing short walls demands considerable trade skill. With ashlar work there is not much problem, as the units have regular and square sides. The stones can be plumbed, squared and levelled, operations which cannot be done easily with rubble work because of the irregularity and unevenness of the units.

To build piers and the like in roughly dressed stones a profile is first erected and braced up firmly at the position of the pier. The profile is made of lengths of timber which are of suitable thickness to avoid them bending out of shape when the stones are being laid. It must be plumb and be of dimensions equal to those of the finished pier.

This method of work should also be adopted in constructing a free-standing stone wall. If a wall is curved a different type of template is required in order to build the wall to the curve. To construct the template, the full size radius of the curve is used and the curve is reproduced on a 4 mm thick plywood. Carefully cut along the line of curve with a padsaw. The template is further strengthened by nailing 50 mm pieces of wood on to its back. To use the template, the straight walls at both sides of the curve are first erected and properly aligned. The template is then placed so that both ends touch the side walls already built, and the stones set to curve. Each course or section of the curve is checked for correctness by placing the template against it. Figure 173 shows the two different kinds of wall profiles and the ways in which they are used.

Finishings to masonry joints

Joints in masonry work can be finished in a number of ways, using any of the traditional joint finishes applied to block wall, i.e. flush, tuck, recessed, or tooled.

Flush joints are usually applied to ashlar work and low class rubble work, such as boundary and garden walls. This type of finish is most suitable for flat surfaced units.

Tuck and recessed joints produce very attractive visual effects and are used for high class masonry work.

Procedure for tuck-pointing

A wall to be tuck-pointed should have the joints raked out as the building proceeds. Pointing is usually started from the upper part of the wall, so that scaffolding can be taken down sectionally and put-log holes made good in the process. Mix mortar in the proportion or 1 part cement to 3 parts fine sand. The mortar should not be too wet, as stone does not absorb moisture as do blocks and bricks. A mortar which is too wet tends to 'weep' and drain over the face of stones. It is common to add to the mortar a colouring pigment which contrasts with the appearance of the stone. The mix for such colouring pigments is normally contained in an accompanying manufacturer's leaflet. In the absence of such specifications, the mix should be 1:1:3.

Using the pointing trowel, fill the joints with mortar and press in well until the mortar sticks out about 6 mm. Smooth the joints and cut the edges clean with the

A Rag rubble

B Square rubble

C Square diagonal rubble

D Random rubble

Fig. 174 Types of rubble

point of the trowel. As the joints sneck round the stones, this work demands a lot of care and good craftsmanship. The width of the joint should not be more than 19 mm. After the initial setting, the joints are gently wiped over with a soft brush.

The procedure for recessed or tooled pointing is similar. The main difference lies in recessing the joints to about 6 mm instead of projecting them. Refer to Chapter 13 on Finishes for block walls on p. 111.

Revision questions

1 What are the advantages of using stones for walling in place of sandcrete blocks?

2 Illustrate with sketches, the following masonry tools:
a) stone punch,
b) mason's hammer,
c) boaster,
d) scratcher.

3 With the aid of sketches, differentiate between the following masonry bonds:
a) flint,
b) square rubble,
c) rag rubble.

4 Define the following masonry terms:
a) hearting,
b) bonder,
c) corner stone.

5 Briefly describe the following joint finishes:
a) tooled joint,
b) recessed joint.

Fig. 177 Square rubble

Fig. 175 Coursed basket weave

Fig. 176 Random rubble

Fig. 178 Stone gate pillar in rubble work. Note bonder stones at quoin

13. Finishes for block walls

All walls that are constructed using a wet system require a surface finish of one type or another. These wet constructed walls consist of blocks, bricks or stones laid with mortar and also cast-in-situ concrete walls.

A wall finish may consist of painting, plastering and rendering, pointing the joints, tyrolean application or dry lining with plaster boards.

Purpose of a wall finish

No matter what type of material is used or which method is adopted for a wall finish, the purposes tend to be common for all. Generally, these include the following.

1 Provide resistance to moisture penetration.
2 Give a more pleasant appearance.
3 Protect the structural wall from driving rain and sand storms.
4 Increase the fire resistance, sound and thermal insulation properties of the wall.

The most common types of finishes are discussed below.

Plastering and rendering

There is a controversy about the meanings of the terms plastering and rendering and a number of the definitions given by various writers serve only to confuse the meanings the more. For instance rendering has been described as one of the undercoats of internal plastering. In this book, however, the term plastering is intended to mean the application of wet mixes of fine cementing material and sand to the internal surfaces of a wall, although in certain cases the cementing material is used without the addition of sand. Rendering means the application of this mixture on the external surfaces of walls.

Both applications serve identical purposes, except that while rendering is intended primarily to prevent the penetration of water into the wall, plastering is intended primarily to provide a flat even surface to the wall.

Plastering

Plaster is a traditional internal finish. The manner in which a wall is to be plastered, i.e. the mixes and the number of coats to be applied, is always specified. If plastering is to be given out as a subcontract, the contractor should be brought in at an early stage. It is a good practice, and advisable too, to plaster a small portion of wall on the site using the specified mixes and number of coats. This specimen is constantly observed to give an idea of how the finished work will eventually perform. The binding material is usually ordinary Portland cement and lime.

Mixes

Plastering mixes are specified according to the background wall to be plastered. They should be as close as possible to the background in strength in order to achieve a good bond. The mix commonly used consists of 1 part cement to 1 part lime to 6 parts sand. As an alternative to using lime, a 1 part cement to 1 part plasticiser to 5 parts sand is used. In either case, the sand used must be clean durable and well graded and should conform to the requirements of B.S. 882: 1983.

The preparation of mixes is the same as described for mortar, or alternatively B.S. 5628: Part 1: 1978 should be consulted.

Preparing the surface to be plastered or rendered

The preparation of the surface to be plastered is necessary in order to ensure a good bond between the surface and the plaster. The surfaces of concrete blocks are usually sufficiently rough to provide a good key and no special preparation is necessary. If however the blocks are dense and close textured, the joints should be raked to a depth of

Fig. 179 Preparing the under surface of a reinforced concrete floor slab to receive plastering at a later stage, using cement grout to form a rough key on drying

13 mm during the process of laying them. Old work may need hacking and cleaning down before the plastering coats are applied. A spatter-dash coat is often applied to dense blocks and to walls that absorb moisture quickly. Minor irregularities, such as crevices, deformations on blocks, open joints, etc. are first made good.

Method of application
The method of applying the mortar mix to the wall is by using the plastering trowel or steel float to collect the mortar from the hawk and pressing it on to the wall in a sweeping stroke. This spreads the mortar in an almost even thickness. Although this is the traditional way of applying the mortar on to the wall surface, several other methods have evolved among local plasterers. These mainly consist of throwing the mortar to the wall surface using the blocklaying trowel. The mortar is picked up from the mortar board and the plasterer, standing sideways to the wall, throws the mortar with an upward stroke to the wall. It is quite a fascinating method to watch and the dexterity with which the experienced plasterer applies the mortar is amazing to the hawk and float user. Each method certainly has its merits and demerits.

Procedure for plastering
1 Clean the surface thoroughly.
2 Spray the background with water and apply cement slurry spatter-dash.
3 Apply the first coat of plaster.
4 Roughen the first coat using a metal comb or lath-scratcher to provide a good key for the second coat.
5 Cure the first coat for at least 48 hours.
6 Place plaster screeds at convenient distances to form guides for straightening the surface. Screeds are simply narrow bands of plastering mortar formed by bedding short pieces of lath in plaster dots. These dots are plumbed down and intermediate ones fixed. The spaces in between the dots are then connected with plastering material to form horizontal or vertical bands about 75 mm wide.
7 Flank-in the spaces between the screeds and rule off with a straight-edge to a straight and even surface.
8 Use the wooden float to work the surface to a sandy gritty finish. For rendering, this forms the final finish.
9 Clean up the working place.

Rendering

The procedure for rendering is much the same as has been described for plastering. We shall, therefore, devote the space to discussing the common problems that arise from rendering external cement-based walls, especially sandcrete block walls.

Problems arising

Cracks

Cracks occur in rendering either because of movement in the rendering itself or because of movement in the background walls. These movements may be in the form of shrinkage. In the case of rendering, most shrinkage cracks are due to cement mixes in the rendering mortar which are too rich. The persistent warmth from the sun's radiation evaporates water from the mortar very quickly soon after mixing. This rather too rapid withdrawal of moisture causes the rich cement mortar to shrink leading to ugly cracks in the rendering. It is therefore a good practice to wet the surface to be rendered well before commencing work.

Adequate adhesion is also an essential precaution against cracking. The mortar mix should not be made richer in cement than the wall to be rendered and, similarly, finishing coats should not normally be stronger than undercoats.

Workmanship is also a factor that contributes to cracking. It is often noticed that workers, in a bid to finish their allocation of work for the day and possibly be able to do some additional bonus work (overtime), trowel a wet finishing coat. In doing so, the cement content of the mix is attracted to the surface, causing high drying shrinkage. All rendering coats must be left to dry before a subsequent coat is applied, and finishing coats must not be trowelled wet. A good rule to use is when a touch of the finishing coat does not leave water on the palm, the rendering is ready for steel trowelling. On no account should clean cement powder or slurry be sprinkled on the finishing coat before trowelling. This practice is often common but it is wrong and must not be encouraged. Loamy and soft sands also shrink badly on drying.

The drying effect of strong wind can be another cause of cracking in rendering. Constant wetting of the work is therefore important to achieve a reasonable standard of finish during windy conditions.

Cracks due to shrinkage of the background are mostly caused by insufficient drying out before the rendering is done. Shrinkage in sandcrete blocks is greater than in any other kind of conventional material. The need for blocks to dry out properly before being used has already been stressed. No sandcrete block walls should be rendered until at least one month after the date of manufacture of the blocks.

Bond failure

This problem is recognised by the presence of hollow patches, flaking of top coats, and peeling of substantial areas of rendering. Although persistent moisture penetration from or through the background will almost certainly result in bond failure, the causes of this problem in rendering work can lie elsewhere. A wall of smooth dense blocks may not offer enough mechanical key and may require some additional treatment to ensure a good bond. The porosity and suction of the background affect adhesion of rendering and its utmost strength. The precautions stated above, such as spraying the background with water before rendering, are essential, particularly during windy conditions when wall surfaces can become covered with a film of fine dust.

Crazing or hair cracks

If the wall is to be painted after rendering, hair cracks or crazing, which are tiny cracks that appear on newly completed rendering, do not pose any serious danger to the surface finish. The cracks are so tiny that coats of paint can easily cover them without trace. Such crazing may be caused by a high suction of the background or undercoat. A sprinkle of water before rendering will prevent premature drying-out of the rendering, and is always advisable. The use of clean cement in powder or slurry form to daub the surface of a finishing coat must be avoided, as this increases crazing. One should protect against excessive direct sunshine and wind during the early hardening period by frequently spraying with water or by covering the new work with polythene sheeting or similar sheeting material.

Single coat rendering increases the possibility of crazing. There should be at least two coats.

All rendering must be adequately cured. A choice can be made between the more traditional method of spraying with water and using curing compounds. Curing compounds can be obtained colourless or slightly coloured and are very economical to use.

Tyrolean finish

There are other methods of finishing block walls besides plastering and rendering. Tyrolean is one very popular finish and has a rough pleasing appearance. It may be coloured according to taste or applied in the natural grey colour of ordinary Portland cement.

A wall which is to receive tyrolean finish is first of all rendered with one or more coats, depending upon how flat and even the wall was built. This is done in order to obtain an even thickness for the tyrolean finish. If the wall is fairly flat and unrendered, the joints must be filled flush and two or three coats of tyrolean applied. The first coat consists of 1 part waterproof cement, and 2–3 parts sand. The second coat consists of 1 part cement, 1 part lime, and 8 parts sand, while the finishing coat consists of 1 part coloured cement, such as Snowcem or Cullamix, and 2½–3 parts sand. The mix is applied with a special hand machine which flips the mixture evenly on the surface of the wall. (See Fig. 180.)

The finish is recommended for public buildings such as schools, hospitals, offices and all block walls in densely populated areas, for the following reasons.

1 Does not require periodic renewal.
2 Driving rain and sand storms do not affect the surface texture.
3 The rough, ragged appearance discourages vandals and children from defacing the wall.
4 Fairly easy to apply and cheap if ordinary cement is used. There is however a disadvantage with its use. The wall can easily be climbed by geckos, lizards and rodents. Lizards can easily deface a wall with their waste which in this case would be difficult to remove. This should be taken into consideration as lizards are plentiful in most tropical areas.

Fig. 180 A good example of tyrolean finish applied to external surfaces of concrete block walls. It is a suitable finish for walls of school buildings

Wall tiling

Tiles are normally used as interior wall finishes. The tiles are made of earthenware, pressed in steel dies, fired in a kiln and glazed during a second firing. They used generally to be white in colour, but various colours can now be obtained. Glazed tiles are usually from 6 mm to 10 mm thick and 100 mm square. The laying is done after the plastering coats have been applied. The major problem in tiling work is bond failure. There must be adequate mechanical key provided by the background and therefore the plaster should be left unsmoothed.

The unglazed back of the tile is porous and absorbs moisture very fast. It is therefore advisable to soak tiles in water before fixing. Tiles should not be fixed on thin backgrounds unless they are made adequately rigid, as tiles on such backgrounds are liable to come loose. Tiles tend to be very fragile and difficult to cut by manual means. The work must therefore be carefully set out before the fixing operation is commenced.

Fixing procedure

1 Hack the wall and clean thoroughly if surface is smooth.
2 Set out the work, starting from a corner, making sure that all odd pieces are placed at a corner and at the base of the wall.
3 Bed a flat piece of lath to a perfect level at the base of the wall to receive the first course of tiles. This may not be necessary with a finished floor.
4 Clean down the wall to remove dust, and spray water on it.
5 Soak the tiles in water.
6 Mix mortar: 1 part cement to 3 parts sand. Fixing of tiles is nowadays done with special mastics.
7 Butter the tile bed evenly with mortar

and press on to the wall, tapping it slightly into position. Ensure that the mortar or mastic covers the entire bed of the tile. The practice of dotting the bed of the tile with mortar or mastic should not be employed as it eventually leads to bond failure.

8 Allow as small a joint as possible between tiles and ensure plumb and flatness by using a spirit level and a straight-edge. By moving the face of the fingers over the edges of newly fixed tiles, one can easily tell if the edges are flat to each other or not.

9 Rub the joints flush with a piece of soft waste cloth.

10 Clean the tiles after setting and polish as a final operation.

11 Clean the tools and working area.

Pointing and jointing

Pointing and jointing have been described briefly in Chapter 12. However it will be necessary here to describe these applications in greater detail.

Pointing

This is the name given to the method of finish the wall joints receive after the wall has been built. There are three types of pointing, each of which is carried out after the joints have been raked out and thoroughly cleaned. These are: weather pointing, weather struck joint and tuck pointing. The procedure for pointing is as described on p. 105 and the reader should refer to it. (See also Fig. 181.)

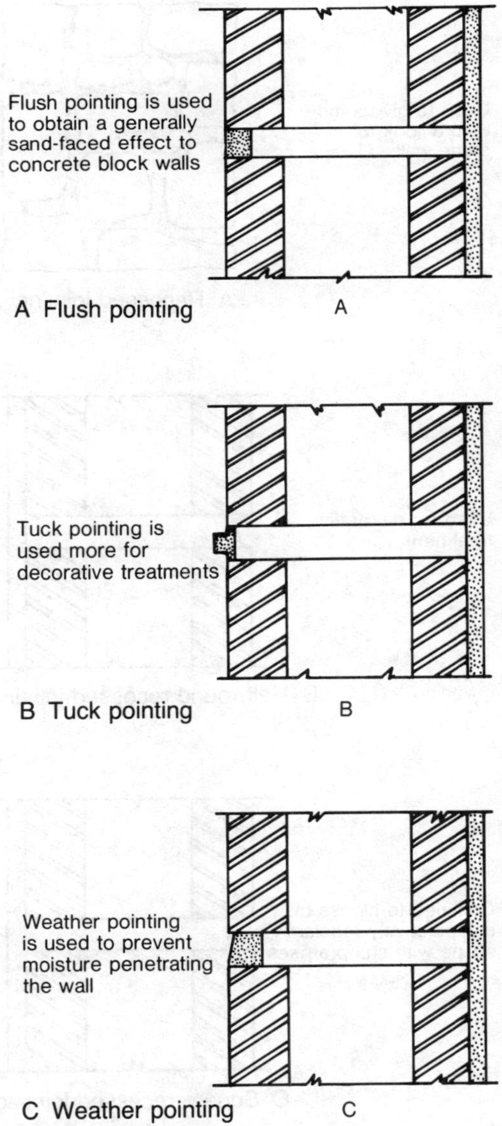

Fig. 181 Types of pointing

Jointing

This is the name given to the treatment of the joints as the building of the wall proceeds. The types of jointing include flush, tooled and recessed. The advantage of jointing is that one worker does the job and so the work is faster and an even colour is obtained.

Procedure for flush jointing

After a reasonable number of courses or portion of block wall have been laid and corrected, i.e. levelled, plumbed and aligned, a piece of clean flat wood is rubbed over the fresh joints. While the rub is being used, joints with insufficient mortar are filled, so that truly flush joints can be obtained. Portions should be finished soon after laying to avoid drying out. In hot weather mortar dries out very quickly and it is not advisable to leave joints unfinished for too long.

If the joints are to be finished recessed, after the rubbing with flat wood, a round metal rod jointer is used by pressing the jointer against the fresh mortar and drawing it to and fro. A spirit level is held against the joint to ensure that the recess is perfectly horizontal and perpends are true.

The finished work is wiped with a soft brush to remove all pieces of mortar drops. Different types of jointing in common use are shown in Fig. 182.

Fixing pre-cast concrete units

Thin pre-cast concrete slabs ranging from

50 mm to 100 mm thick, depending upon the area to be covered by a slab, may be fixed to the face of a block wall to achieve an architectural effect. When a pre-cast concrete slab is fixed to the wall surface, the work is referred to as cladding. The general aims of cladding include providing a weather proof surface for the wall by using light, suitable materials and giving a surface finish that is attractive.

A variety of surface textures can be obtained using concrete as the backing material. Reconstructed stone resembles natural stone and is obtained by the use of granite stone dust as the aggregate and cement. The cement can be of any suitable colour. Exposed aggregate finish is another popular way of making pre-cast concrete slab units; this may be achieved by grinding off the slab surface as is done in terrazzo works or rolling selected aggregates into the slab while it is still green. Another method is to lay the selected aggregates face down on a thin layer of sand in the mould. The concrete, which is mixed semi-dry, is then laid on top and compacted. The sand is later washed out when the slab has set and hardened. Such a variety of surface finish is unobtainable with traditional materials.

Fixing and jointing

The slabs are fixed to the background wall by means of non-ferrous metal cramps. These are adequately bonded into the slab during the casting operation and the ends built into the backing wall when fixing. The fixing should be done in such a way that the slabs and the backing wall can move independently.

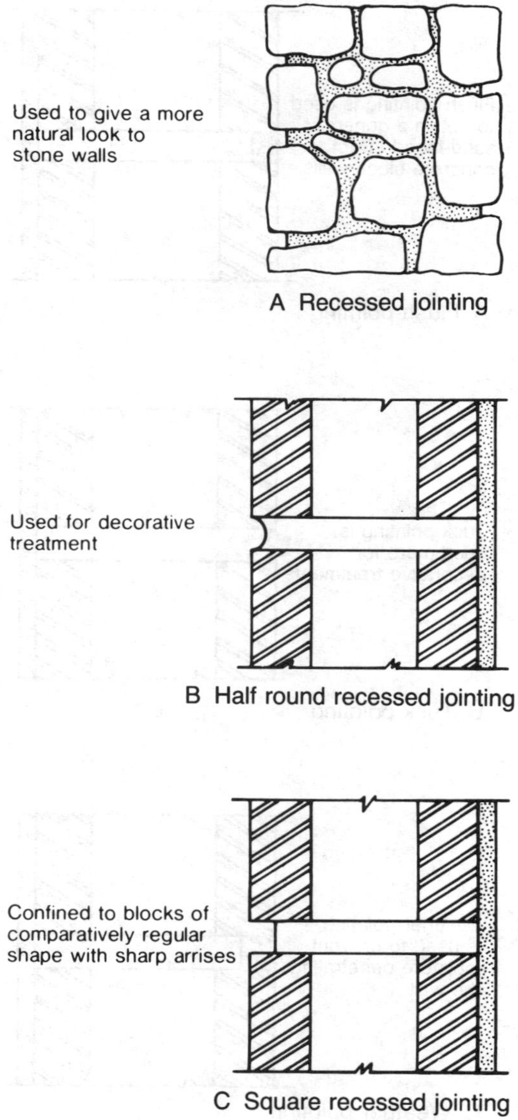

Fig. 182 Types of jointing

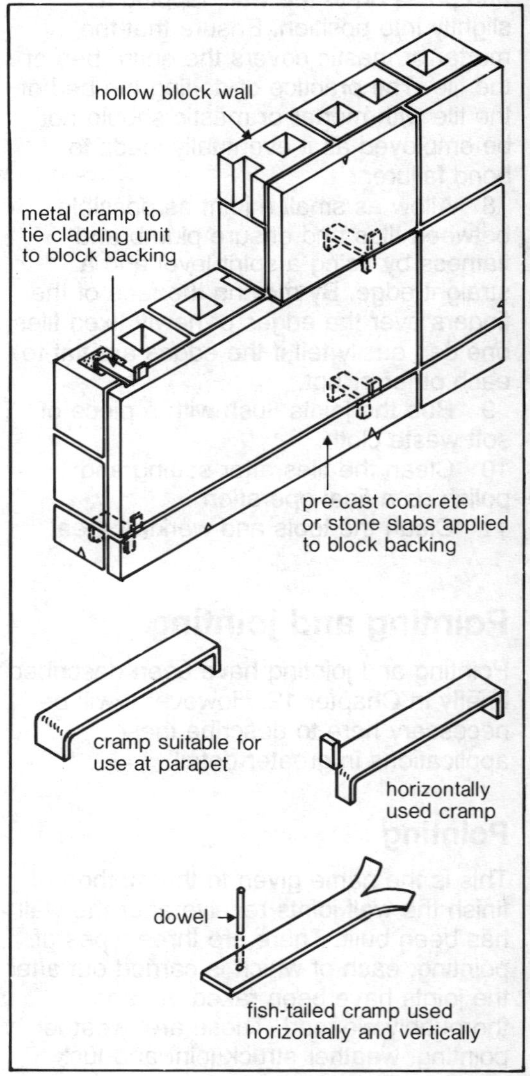

Fig. 183 Fixing details for cladding

Because the slabs are exposed to direct rays of the sun, they are affected by temperature changes. Expansion joints are therefore necessary to accommodate the possible differential thermal movements. These joints are determined by the type of design of the building but 6·5 m intervals is a usual distance between joints in concrete slab cladding. The jointing material should be unaffected by changes in the temperature and the weather condition; bituminised foamed polyurethane or soft mastic are both suitable materials. The width of the joint should not exceed 5 mm. Some fixing details are illustrated in Fig. 183.

Revision questions

1 State the main purpose of a wall finish.
2 Define the terms plastering and rendering.
3 What are the precautions to be taken when plastering a wall to ensure that the finished work does not flake off?
4 Describe a procedure for plastering a block wall.
5 a) What are likely causes of cracking in plastering finishes?
b) Desribe two other common failures in plastering and suggest ways of remedying them.
6 For what kinds of work are the following finishes most appropriate:
a) tyrolean,
b) glazed ceramic tiles,
c) cement/sand plaster,
d) flush jointing?
7 Describe, with the aid of sketches, a method of fixing a pre-cast concrete cladding slab to a block wall.

14. Scaffolding

A scaffold is a temporary platform erected to enable building operatives to gain easy and safe access to their work above ground level. It is also used to carry the materials and tools necessary for the execution of the work.

Scaffolding can be classfied in either of two ways:

1 by the material used for the scaffold, i.e. wood or steel or light alloy tubes,

2 by the method of erection, i.e. dependent or independent.

Each of these is discussed in greater detail later, but first the common terms and parts of a scaffold will be looked at.

Parts and terms used in scaffolding

(Refer to Figs. 184 and 185.)

Standard
The upright members positioned in alignment and parallel to the wall face.

Ledgers
The horizontal members which connect the row of standards.

Putlogs
Shorter lengths with one end flattened to enter the wall joints.

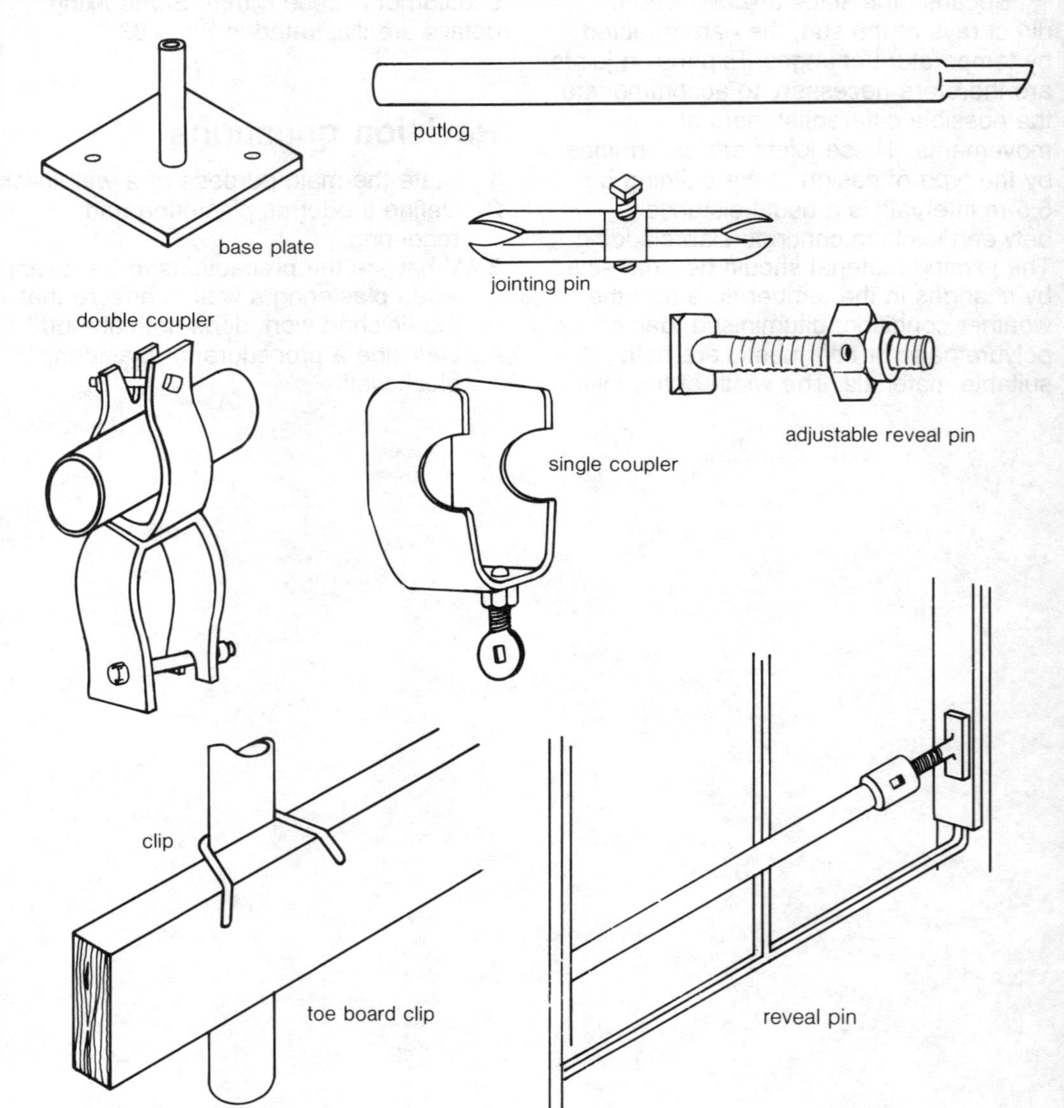

Fig. 184 Scaffolding fittings and parts

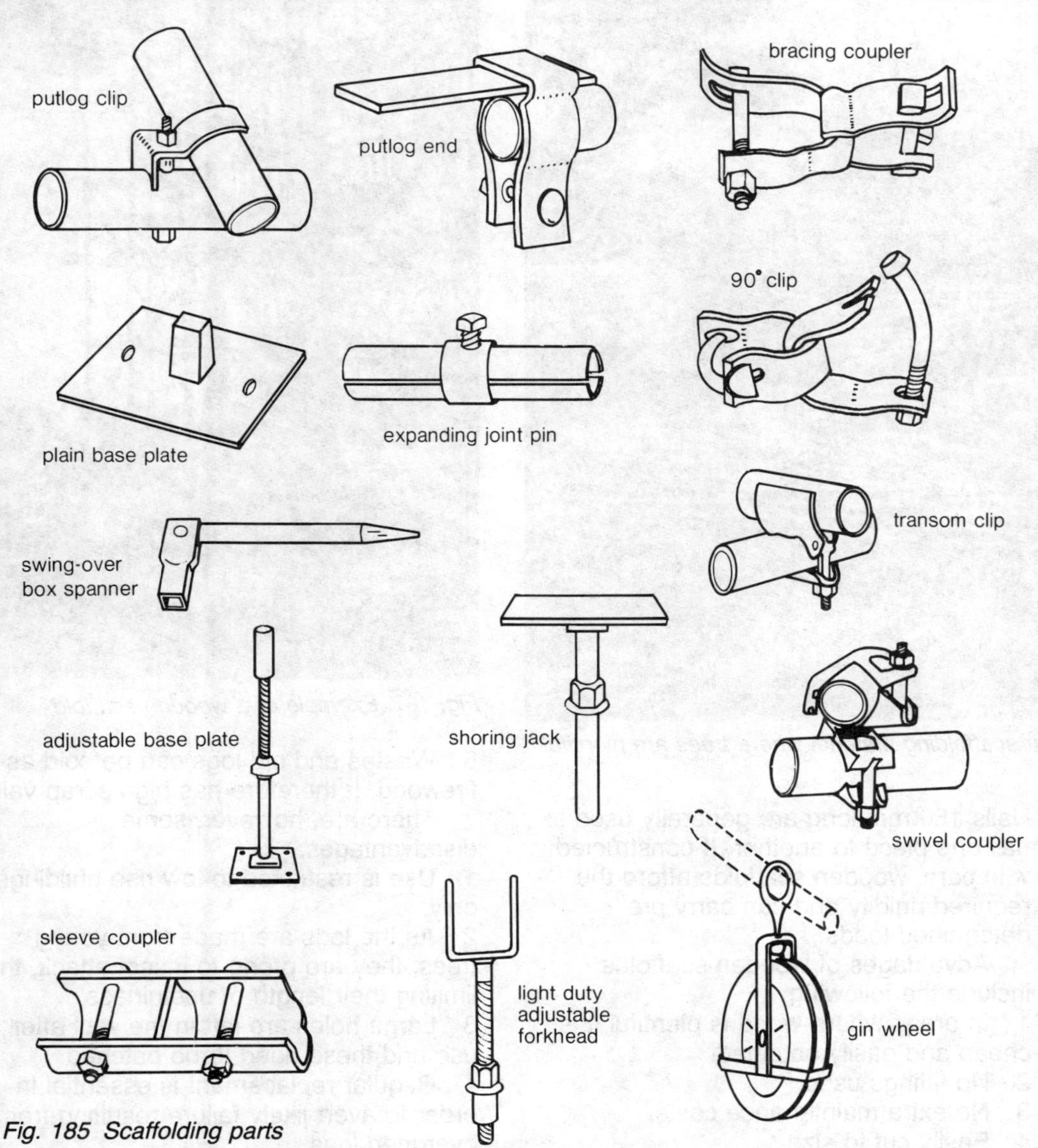

Fig. 185 Scaffolding parts

Brace
A ledger fixed diagonally across the standards for added rigidity.

Reveal pin
A short piece with screwable end attachments used across openings for the purpose of fixing a putlog.

Base plate
A base plate is used at the foot of the standard to provide a stable footing. The base is nailed to a base board if the ground upon which the scaffold rests is weak and unstable.

Double coupling or coupler
This is used to tie ledgers to standards at right angles.

Single coupler or putlog clip
Used to tie a putlog to a ledger.

Swivel coupler
A rotating fitting used to tie a brace to the standard.

Jointing pin
Used to join two lengths of tubing. The pin consists of split halves connected by a coach screw. Tightening the bolt causes the split halves to pull apart thereby wedging tightly against the internal surface of the tubes.

Classification by type of material

As briefly pointed out above, there are two materials that may be used in scaffolding: wood and metal.

Fig. 186 Stacks of logs of wood. Wood is a popular scaffolding material where trees are plentiful

Fig. 187 Example of a wooden scaffold

Wooden scaffold

Wood is commonly used for scaffolding, mostly as props for concrete decking forms. For most single storey blocks, wooden scaffolds are used to great advantage. The logs are obtained from young fairly straight trees which are cut to about 4 m lengths. The branches are trimmed off and the bark cleaned away. This helps the logs to dry quickly. They are used while they still retain enough sap to make them nailable and elastic.

Nails 150 mm long are generally used to nail one piece to another. If constructed with care, wooden scaffolds afford the required rigidity and can carry pre-determined loads.

Advantages of wooden scaffolds include the following.

1 In areas where wood is plentiful it is cheap and easily obtained.
2 No fittings used.
3 No extra maintenance costs.
4 Easily cut to size.
5 Wastes and old logs can be sold as firewood. It therefore has high scrap value.

There are, however, some disadvantages.

1 Use is restricted to low rise buildings only.
2 As the logs are made from young trees, they are prone to insect attack, thus limiting their length of usefulness.
3 Large holes are left in the wall after use and these need to be patched.
4 Regular replacement is essential in order to avert likely failure resulting from overdried logs.

Fig. 188 Scaffold parts stacked on site according to types and lengths

Tubular scaffold

Steel and light-alloy tubes are the commonest materials used for this type of scaffolding although the use of steel is fast diminishing because it is heavy and costly to maintain. The scaffold is erected using a series of fittings.

Advantages of tubular scaffolding include the following.

1 Less likely to deteriorate than wooden scaffolding.
2 Correctly used it is more rapidly erected.
3 More convenient for internal work owing to wider range of widths.
4 Takes up less space when stored.
5 Used for multi-storey buildings.

Disadvantages are as follows.

1 Cost of preservation.
2 High initial cost.
3 Many types of couplers needed.

Regulations affecting scaffolds

The Construction (Working Places) Regulations 1966, are chiefly concerned with scaffolds and similar structures and appliances from which building operations are carried out. The regulations are very detailed and deal with the quality of materials, erection, inspection, usage and maintenance.

Falls from and incorrect use of scaffolds account for the majority of accidents on the construction site. It is therefore, our professional responsibility to see that regulations guiding scaffolds are strictly obeyed.

The following summarises the contents of the Construction (Working Places) Regulations 1966, in respect of scaffolds.

Scaffolds and components used for them must be of sound materials, of safe construction and be sufficiently strong to serve the purpose for which they are meant.

A partly erected or dismantled scaffold must comply with these regulations and if not must not be allowed to remain in that condition unless a warning notice forbidding its use is displayed prominently at its approaches. Alternatively the approaches to it must be effectively locked.

Timber members of scaffolds, ladders and trestles must not be treated in such a way as to hide defects on them. Metal parts must be free from corrosion and similar defects.

The standards of a scaffold must either be vertical or lean slightly towards the building for stability.

The foot or base must not be placed on loose or weak material, but placed on an adequate base plate so as to prevent slipping or sinking.

Putlogs or other supports for a platform must be securely fastened to the ledger or standard or their movement restrained by other means.

If a putlog is supported by a wall, it must extend into the wall to provide an adequate support.

The distance between two consecutive putlogs varies according to the thickness of planks used for the platform.

1 For 32 mm thick planks, it must not exceed 1 m.
2 For 38 mm thick planks, it must not exceed 1·5 m.
3 For 50 mm thick planks, it must not exceed 2·6 mm.

The minimum width of working platforms, from which a person is liable to fall a distance of 2 m or more, must be as follows.

1 As a footing only, minimum width must be 600 mm (3 boards).
2 For deposit of material, minimum width must be 800 mm (4 boards).
3 For passage of materials, minimum width must be 600 mm (3 boards).
4 For dressing of stone, minimum width must be 1·30 m (6 boards).
5 For support of higher platform, minimum width must be 1·05 m (5 boards).

6 For higher platform and stone dressing, minimum width must be 1·50 m (7 boards).

Planks should not project beyond the end support at the end of the scaffold, or at a change of direction, any more than a distance of four times the thickness of the board. The planks must be fixed securely to prevent tipping.

Provision must be made for levelling pieces where boards overlap to make the movement of wheelbarrows easier and to prevent falls.

A clear passage of at least 400 mm must be left between rails of the working platform and any obstruction or deposited material.

Braces must be used to prevent the scaffold from collapsing due to sideways sway. Where the scaffold is adequately tied to the structure at its upper end, no braces need be used.

Every scaffold must be tied to the structure vertically and horizontally every 10 m.

The distance between two consecutive standards in a normal bricklayer's or blocklayer's scaffold should not be more than 2·5 m. Between ledgers, which is termed a 'lift', it should not be higher than 1·6 m.

Guard rails and toe boards must be provided in all situations where persons are liable to fall a distance of 2 m or more.

A guard rail or rails must be fixed 910 mm from the top of the toe board or at 1·15 m from the working platform.

The minimum height for a toe board is 150 mm and the distance between toe board and the lowest guard rail must not exceed 765 mm. Guard rails and toe boards must be placed on the inside of the standards.

Guard rails and toe boards are not required for the platform of a ladder scaffold if there is a secure hand-hold for the full length, nor for the platform of a trestle scaffold which is supported by the trestle.

The use of a trestle scaffold is not allowed if persons are liable to fall more than 4·5 m from the platform which it supports. A trestle scaffold must not be constructed with more than one tier if folding supports are used.

Ladders are often used on scaffolds, and the following regulations relate to them. Ladders of all kinds other than those used as crawling devices on roofs must stand on a firm base. They must be securely fixed, near their upper resting place when in use and, if this is impracticable, at or near their lower ends. Where neither fixing is practicable, an adult must be stationed at the foot to prevent slipping. Ladders must have a firm base and not stand on loose blocks or other loose materials.

Methods of erection

There are two ways in which scaffolds, other than trestle scaffolds, may be erected.

Dependent scaffold

This is variously termed a 'putlog scaffold' and 'bricklayer's scaffold'. The system uses putlogs whose flat ends are built into the wall joints to provide rigidity. The stability of dependent scaffolds is provided by these putlogs. There is only one row of standards connected by ledgers. The erection procedure is given below. Figure 189 shows a dependent scaffold.

Independent scaffold

This is sometimes called the 'mason's scaffold'. It has two rows of standards, one close to the structure and another at a distance equal to the required platform width away from the first. It is stronger, and is used for heavy operations, such as masonry work. Putlogs are not used; instead transom pieces are used to support the platform (see Fig. 190).

Procedure for erecting a tubular scaffold

The scaffolding units have been brought to the site and sorted out in their various lengths, fittings oiled and sorted out according to their kinds, and planks stacked ready.

1 Place the base plate for standards No. 1 and No. 3 at equal distances from the wall, allowing for the type of platform.

2 With the aid of an assistant, stand the standards on the base plates and connect the two with a ledger using double couplers. The ledger should be sighted

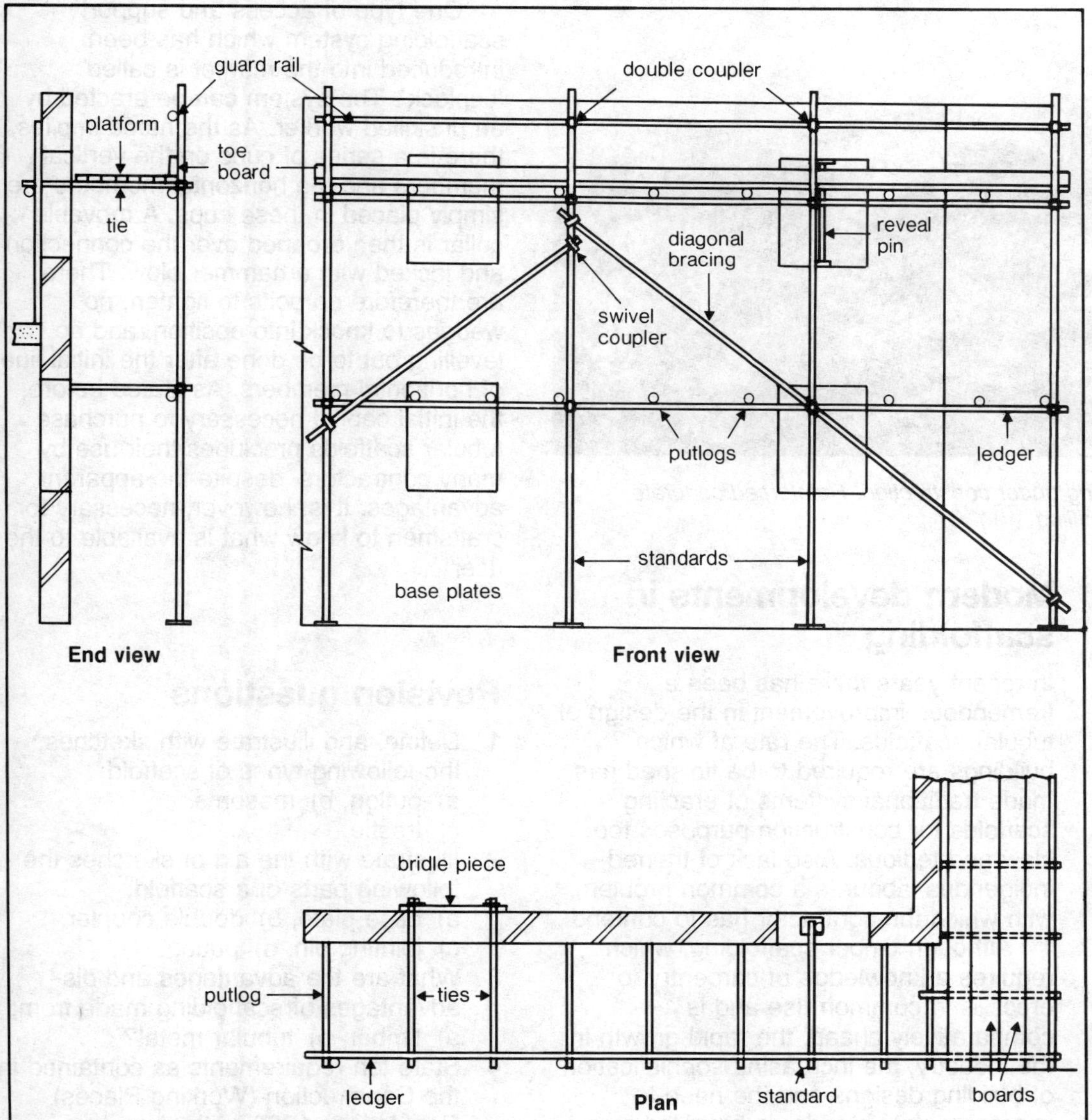

Fig. 189 Putlog or dependent scaffold

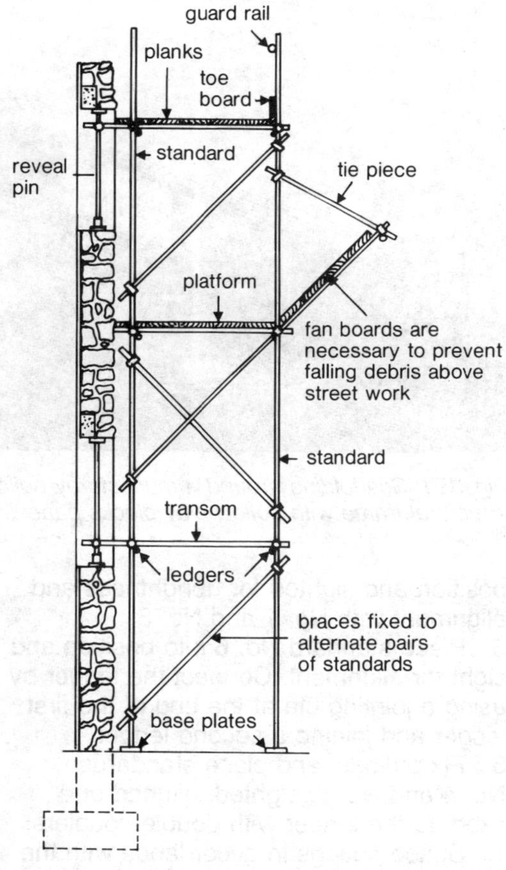

Fig. 190 Mason's or independent scaffold

for level. The bolts of the couplers should be placed uppermost.

3 Putlogs can now be fixed using single couplers or putlog clips with the flat ends pushed well into the joints.

4 No. 2 standard is then placed into

Fig. 191 Scaffolding around a multi-storey building under construction. Reinforced concrete structural frame with hollow clay blocks panel in filling

position and sighted for uprightness and alignment with No. 1 and No. 3.
5 Place standard No. 6 into position and sight for alignment. Connect the ledger by using a jointing pin at the end of the first ledger and joining a second ledger.
6 Fix putlogs, and place standards No. 4 and No. 5, sighted, aligned and fixed, to the ledger with double couplers.
7 Space putlogs in accordance with the regulations, but place two putlogs close together at the junction of planks so as to prevent tipping.
8 Repeat this procedure until the required length of scaffold is obtained. At corners, it is usual that the planks from one direction overlap those from the other and a ledger comes over the other at a right angle. Putlog holes are left out or made in the cross joints of the lower ledger.

Modern developments in scaffolding

In recent years there has been a tremendous improvement in the design of tubular scaffolds. The rate at which buildings are required to be finished has made traditional systems of erecting scaffolds for construction purposes too slow and tedious. Also lack of trained indigenous labour is a common problem with which the contractor has to contend.

Although timber scaffolding, which requires a knowledge of carpentry to erect, is in common use and is comparatively cheap, the rapid growth in the industry, the increasing sophistication of building designs and the need to improve safety standards have led many builders to seek other ways of scaffolding.

One type of access and support scaffolding system which has been introduced into the market is called 'cuplock'. The system can be erected by an unskilled worker. As the name implies, there is a series of cups on the vertical members and the horizontal members are simply placed in these cups. A movable collar is then dropped over the connection and locked with a hammer blow. There are therefore no bolts to tighten, no wedges to knock into position, and no levelling out to be done after the initial line of horizontal members. As stated before, the initial capital necessary to purchase tubular scaffolds precludes their use by many contractors, despite the apparent advantages. It is, however, necessary for craftsmen to know what is available to the user.

Revision questions

1 Define, and illustrate with sketches, the following types of scaffold:
a) putlog, b) mason's,
c) trestle.

2 Illustrate with the aid of sketches the following parts of a scaffold:
a) base plate, b) double coupler,
c) jointing pin, d) putlog.

3 What are the advantages and disadvantages of scaffolding made from:
a) timber, b) tubular metal?

4 State ten requirements as contained in the Construction (Working Places) Regulations 1966, with regard to scaffolds.

15. Drainage

Drainage, as distinct from plumbing, is a system whereby waste products from baths, wash-basins, water closets and sinks together with rain are collected and discharged into public sewers, septic tanks, soakaways or cesspools. The drain layer's work begins where the plumber's stops.

The construction of drains is therefore principally the concern of the blocklayer. The very poor way in which drainage work is carried out in some towns and cities makes it necessary to treat the subject in some depth at this stage of studies. A radical change in housing drainage in the future depends mainly on the seriousness that is attached to the subject now.

Principles of drainage

The following principles apply to every drainage scheme.

1 Materials used to be of the best quality.

2 Drains to be provided with sound foundations laid with uniform gradient (the availability of flexible plastic pipes may make this provision unnecessary).

3 Adequate means of inspection and cleaning to be provided by the construction of inspection chambers at changes of direction and other convenient positions. As many branch drains should be included as possible.

4 Pipes should, as far as it is practicable, be laid in straight lines between inspection chambers.

5 Drains on completion must be airtight and watertight, be of good workmanship with a clear bore throughout and be self-cleansing.

6 All traps must be self-cleansing and have an adequate water seal.

7 Branches should connect the main line in an inspection chamber. In some circumstances, it is possible to use a branch junction which joins the main drain obliquely in the direction of the flow.

8 Drains to have adequate ventilation.

9 Rainwater pipes, sinks, lavatory basins and bath waste must discharge over gullies outside the building.

10 Water closets must connect direct to drains, upper water closets being connected to well ventilated soil pipes.

11 No drain should pass under a building, unless it becomes unavoidable, in which case the section under the building must be encased in concrete not less than 150 mm thick.

Systems of drainage

There are two main systems of drainage.

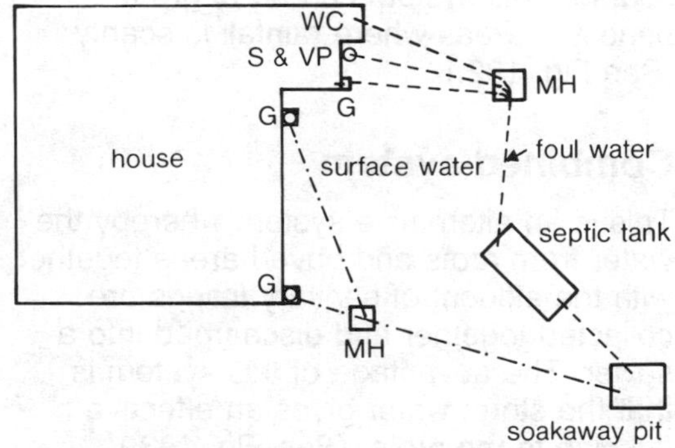

Fig. 192 Separate drainage system

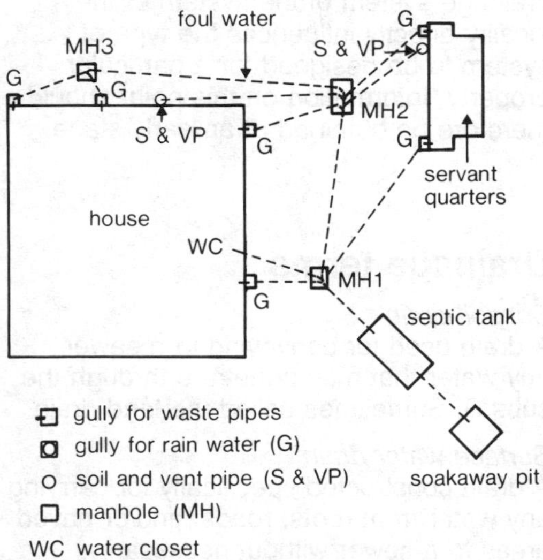

Fig. 193 Combined drainage system

Separate system

In this system, the rainwater is kept separate from the foul water from the house. The rainwater is collected and either discharged into water tanks to be used later or discharged into a water course. This system can be of great benefit in areas where rainfall is scanty. (See Fig. 192.)

Combined system

This is an alternative system whereby the water from roofs and paved areas together with the effluent of sanitary fittings are collected together and discharged into a sewer. The advantage of this system is that the storm water gives an effective flushing to the drain. (See Fig. 193.)

It is important to note that the main drainage system of the system of the locality directly influences the type of system to be designed for a particular property. Information on this point should therefore be obtained at an early stage.

Drainage terms

Subsoil drain
A drain used for conveying to a sewer any water that may penetrate through the subsoil. Sometimes called the land drain.

Surface water drain
A drain constructed specifically for carrying any water from roofs, roads, and/or paved areas to a sewer without necessarily using pipes for such conveyance.

Sewer
Large diameter underground drain to remove water and refuse from towns and cities.

Sewage
This refers to refuse, waste matter, excrement, etc. conveyed in a sewer.

Sewerage
This is a sewage treatment and disposal plant. An efficient sewerage system is essential for public health.

Drainpipes
These may be made of asbestos, pitch fibre, plastic, or stoneware. They are made in standard lengths. Asbestos and stoneware pipes are made with spigot and socket ends to receive cement joints. The usual diameter for house drainage is 100 mm.

Generally drainage pipes can be grouped into rigid and flexible materials. Rigid pipes consist of cast iron, concrete, asbestos cement and glazed stoneware; while flexible pipes are unplasticised polyvinyl chloride (PVC) and pitch fibre. Flexible pipes can accommodate ground movement in underground drainage systems without fracture. The pipes are made in longer lengths with consequently fewer joints, they are easy to lay and lighter in weight. However, a major disadvantage is that they have been found to become compressed under the weight of backfill and may become completely flattened.

Rigid pipes are more advantageous in this instance, but can be damaged by backfilling too unless due care is exercised. They can also fracture easily if there is an uneven bed.

Easy bends
These are introduced into drainage systems to facilitate changes in direction. They are made in various radii Y junctions. When a branch drain joins a main drain without the use of an inspection chamber, a Y junction is used.

Gully traps
These are receptacles placed at or near ground level, into which all waste from sanitary fittings and rainwater discharge. Water closets must not discharge over gully traps. There are different types of gullies, such as 'P' traps and 'Q' traps, and the type used depends upon the depth of the drain. Gully traps are so designed that under normal conditions a water seal is maintained so as to prevent any gas or smell escaping from the drain. (Figure 194 shows some of these parts of drainage systems.)

Waste water
This means used water not contaminated by soil water or trade effluents.

Waste appliance
This is the term used to describe baths, lavatory basins, domestic sinks, drinking fountains, shower trays, wash fountains, washing troughs and wash tubs.

Soil appliance
A term used to describe a water closet (W.C.) or urinal.

Waste pipe
This means a pipe carrying waste water or soil water together with rainwater.

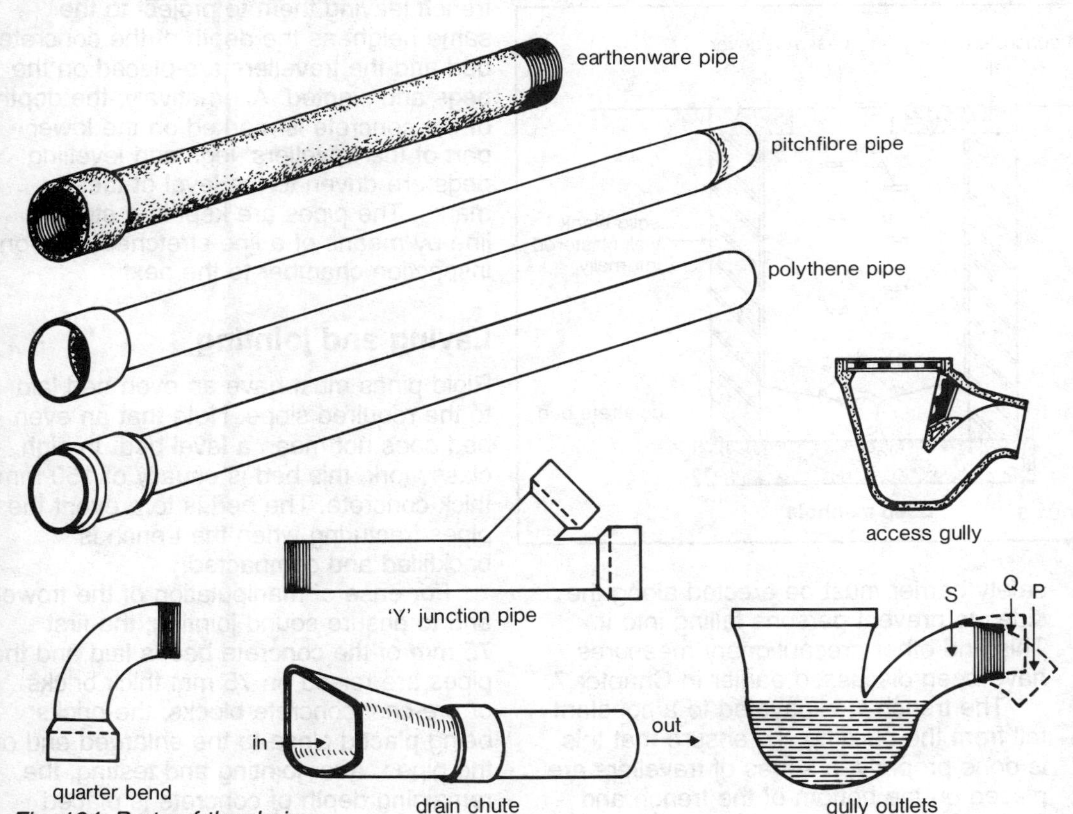

Fig. 194 Parts of the drain

Soil pipe
This is a pipe carrying soil water or carrying soil water together with waste water and/or rainwater.

Rainwater pipe
As the name indicates, this means a pipe carrying rainwater only.

Ventilation pipe
A pipe open to external air at its highest point which ventilates drainage systems either by connection to a drain or soil pipe or waste pipe, and which does not carry soil, waste or rainwater.

Inspection chamber or manhole
This is a chamber constructed on a drain line at convenient distances and at changes of direction for the purpose of access to the drain for inspection or cleansing.

Drain chute
A piece of pipe with an enlarged end, connected to the drain in an inspection chamber to facilitate rodding.

Flaunching or haunching
A method of providing rigidity for the drain by concreting it half way up with a sloped finish.

Benching
A method of protecting half channels in inspection chambers and also of directing overflows of effluent back into the drain. Figure 195 shows a cross-section through an inspection chamber.

Construction of drainage

Setting out and excavation of drainage trenches

The setting out of the trench is much simpler than the setting out for building foundations, because the drain line follows a straight course as much as possible. The two do however follow similar principles. Sight rails, pegs and travellers are needed for the setting out operation. Travellers are T-shaped wooden tools used in a similar way to boning rods.

The centre line of the drainpipe is marked out by wooden pegs driven into the ground starting from the building outlet. A rail, which is a horizontal piece nailed to two vertical members, is erected across the trench at the position of the first manhole or inspection chamber near the building. The height of the rail is fixed at a suitable known level above the invert

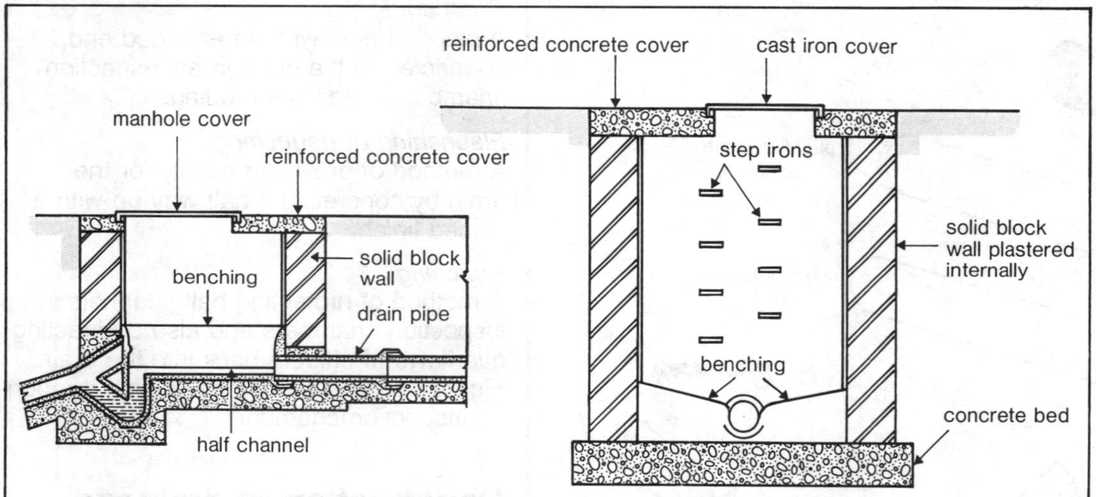

Shallow manhole *Fig. 195 Manholes* **Deep manhole**

level of the pipe. The positions of inspection chambers are then marked out along the line by simply driving four pegs into the ground.

Excavation can commence after all the necessary setting out. Shallow trenches in firm soils up to 1·3 m deep do not need supports unless the trench is close to a road where vibration caused by traffic may cause the collapse of the sides. When a trench is deeper or in weak soil, the sides should be timbered in such a way as to conform to the Building Regulations, to prevent injury to persons. Reference should be made to timbering for trenches discussed in Chapter 7, Trench excavations.

An open drain trench constitutes a potential hazard on the site. When the trench is over 2 m deep a fence or other safety barrier must be erected along the sides to prevent persons falling into it. This and other precautionary measures have been discussed earlier in Chapter 7.

The trench is excavated to a constant fall from the building. To ensure that this is done properly, a series of travellers are placed on the bottom of the trench and the tops sighted to the sight rail. The length of the vertical piece of the traveller is the same as the height of the sight rail measured from the invert level of the pipe. The rails are erected at the inspection chamber positions and at changes of direction. This method gives an acceptable gradient. The sighting of the travellers is done from the lower rail through to the upper one. This method is used for the foundation concrete. In this case levelling pegs are driven into the bottom of the trench leaving them to project to the same height as the depth of the concrete bed and the travellers are placed on the pegs and sighted. Alternatively, the depth of the concrete is marked on the lower part of the travellers' legs and levelling pegs are driven to the level of these marks. The pipes are kept in a straight line by means of a line stretched from one inspection chamber to the next.

Laying and jointing

Rigid pipes must have an even bed laid to the required slope. Note that an even bed does not mean a level bed. In high class work, this bed is usually of 150 mm thick concrete. The bed is to prevent the pipes fracturing when the trench is backfilled and compacted.

For ease of manipulation of the trowel and to ensure sound jointing, the first 75 mm of the concrete bed is laid and the pipes are raised on 75 mm thick bricks or pre-cast concrete blocks, the bricks being placed close to the enlarged end of the pipes. After jointing and testing, the remaining depth of concrete is placed and flaunched.

In less important jobs, a concrete bed may not be necessary, especially if the ground is firm and stable. A well consolidated bed of earth is all that is required, but the backfilling and compacting should be done with a lot of care. The backfilling material should be carefully placed under the pipes and worked solid.

A concrete bed is not required by the regulations for cast iron pipes and, where it is provided, a 100 mm thickness is normally sufficient.

The jointing of pipes made with sockets and spigots is achieved by inserting the spigot into the socket, caulking with tarred hemp or yarn and then making the joint with 1:1 cement and sand mortar. Where this type of joint is used, the drain must not be tested until the cement has gained sufficient strength. This period should be at least twelve hours after jointing. The purpose of caulking the joint with yarn is, first, to centre the spigot in the socket and, second, to prevent mortar falling into the bore of the pipe during the process of jointing.

Cast iron pipes are jointed by inserting a ring of yarn or lead wool into the jointing space and then running in molten lead and caulking. A flexible joint can be obtained by using a rubber ring.

PVC jointing methods use pre-formed sockets or loose couplings, and the joint is made by a rubber ring or using a solvent as an adhesive. Pitch fibre pipes are jointed by means of external couplings in polypropylene. There are two types: one type consists of a coupling which fits over a plain ended pipe and incorporates a rubber sealing ring; the other type is a tapered coupling which is driven on to a pipe with a similarly tapered end. Asbestos cement pipes also have tapered ends which are jointed in much the same way using coupling and rubber rings. The different ways of jointing drainage pipes are illustrated in Fig. 196.

The manufacture of modern stoneware pipes ensures the accurate alignment of the drain and an unobstructed bore for the free passage of the effluent in the form of self-centring pipes, as shown in Fig. 196.

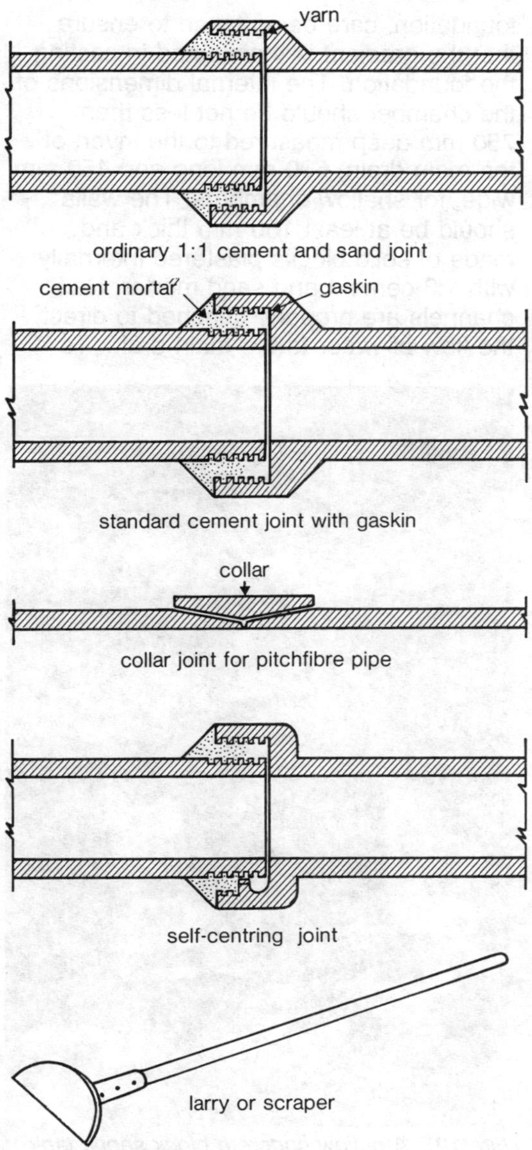

Fig. 196 Jointing drainage pipes

It is absolutely essential for the drain to have no obstructions at all within the bore. This can only be assured at the time of laying the pipes. Obstructions cause blockages in the system, and if they are not removed at the time of laying it is extremely difficult and costly to do so when the system is in use. A tool, called a drain hoe, scraper or larry, is used to remove any mortar that might have found its way into the drain. After the jointing has been completed the tool is passed through the pipe, first along the top and then along the bottom of the pipe, thereby removing any surplus mortar as it is withdrawn through the open end of the pipe. It is a good precaution to further flush each section of the drain with water.

A ball slightly smaller in diameter is dropped at the upper end of the drain and it should roll through to the lower end of each section without stopping. This test checks both that there is no obstruction and that the drain has the correct gradient.

Inspection chambers

Inspection chambers must be provided along the drain line to facilitate access and ensure adequate inspection and cleansing of the drain. Reference should be made to Building Regulations, Part H: 1985 and the associated Approved Document H for further information.

Inspection chambers must be provided:

1 at any change of direction or gradient which would cause difficulty in cleaning a drain or private sewer if no inspection were provided;
2 on a drain line, within 12·5 m of any junction between that drain and any other drain sewer, unless there is an inspection chamber at the junction;
3 at the highest point of any private sewer, unless a rodding-eye is placed at that point.

No part of a drain or private sewer may be more than 45 m (measured along the drain line) from an inspection chamber placed on the same drain or private sewer.

Any inspection chamber must:
1 be designed and built of brickwork or blockwork, concrete or any other suitable and durable materials, and a) be able to withstand all loads to which it may be subjected, b) exclude subsoil water, c) be watertight;
2 be of suitable dimensions and proportions to allow easy access to drain or sewer or private sewer for inspection, cleansing and rodding;
3 have a strong, removable, non-ventilating cover of suitable, durable material;
4 be fitted with step irons, ladders, etc., as necessary to ensure safe access to the drain or private sewer;
5 if the drains in the chamber run in open channels, have a smooth impervious surface benching directing the flow into the main channel, and providing a safe foothold.

To satisfy these conditions, the chamber is built on a solid concrete foundation, care being taken to ensure that the gradient is maintained in casting the foundation. The internal dimensions of the chamber should be not less than 750 mm deep measured to the invert of the main drain, 600 mm long and 450 mm wide, for shallow chambers. The walls should be at least 150 mm thick and made of solid blocks plastered internally with 1:3 cement and sand mortar. Half channels are properly benched to direct the flow of water to the main drain. To assist the flow from branch drains, three-quarter or baron bends are used (Fig. 195). The chamber is covered with either a suitable reinforced pre-cast cover with attached handles or with a patent iron inspection chamber cover. The chamber next to the sewer must have an intercepting trap with a rodding arm, to prevent foul air and gas escaping from the sewer. A ventilation pipe with inbuilt mica flap should also be connected to the chamber. Deep inspection chambers must have strong step irons as needed to facilitate safe access.

Fig. 197 A hollow concrete block septic tank under construction

Septic tanks

A septic tank is an apparatus for sewage purification. It is used in places where a private or public sewer does not exist and is common in many developing countries. The septic tank is a chamber divided into two sections with a receiving, intercepting chamber attached. The first chamber receives the sewage which is later decayed or decomposed by the action of anaerobic bacteria. These bacteria liquefy the organic solids. The liquefied sewage passes on to the filter and aeration section, where the dissolved impurities are removed by the action of aerobic bacteria. Finally the purified effluent is absorbed into the soil. The sludge which settles at the bottom of the septic tank (chamber) is cleaned periodically by mechanical means. Septic tanks are generally designed to treat a pre-determined number of litres per day per person. (See Fig. 198.)

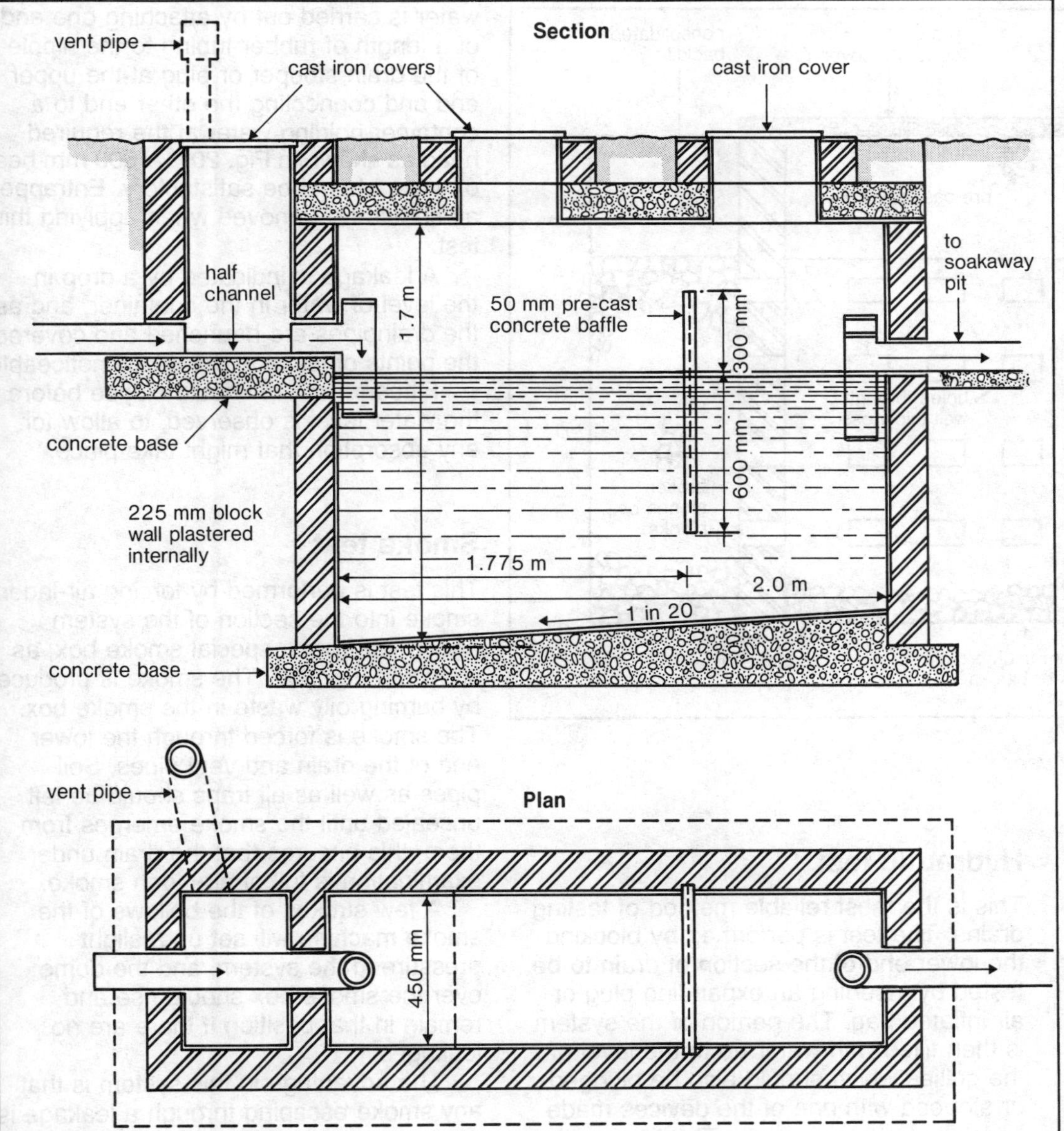

Fig. 198 A typical septic tank

Soakaways

In a separate system of drainage, a soakaway is used to dispose of surface water which cannot be directed into some natural water course. The purified water from the septic tank installation may also run into a soakaway.

The soakaway is a pit dug in a convenient position in permeable soil so that the water that comes into it is absorbed into the soil. Generally, the soakaway is filled with large broken stones and may be covered with a concrete slab to support the sides of the pit and exclude soil; or simply covered with excavated material. It is important to locate the soakaway in a position where the water will seep away, so as to prevent a waterlogged area within close proximity to the building. Stagnant water breeds mosquitoes and this must be avoided at all cost. (See Fig. 199.)

Connection of a drain to a sewer

At the present stage of development of some tropical countries, health plans in towns and cities do not include the construction of public sewers, hence septic tanks are in general use. Where sewers do exist, the connection of a drain must be done by or under the supervision of the local authority. Such drains must be connected to the sewer at the upper half, and in the direction of the flow. A special fitting known as a saddle is used, being bedded into the opening with cement and

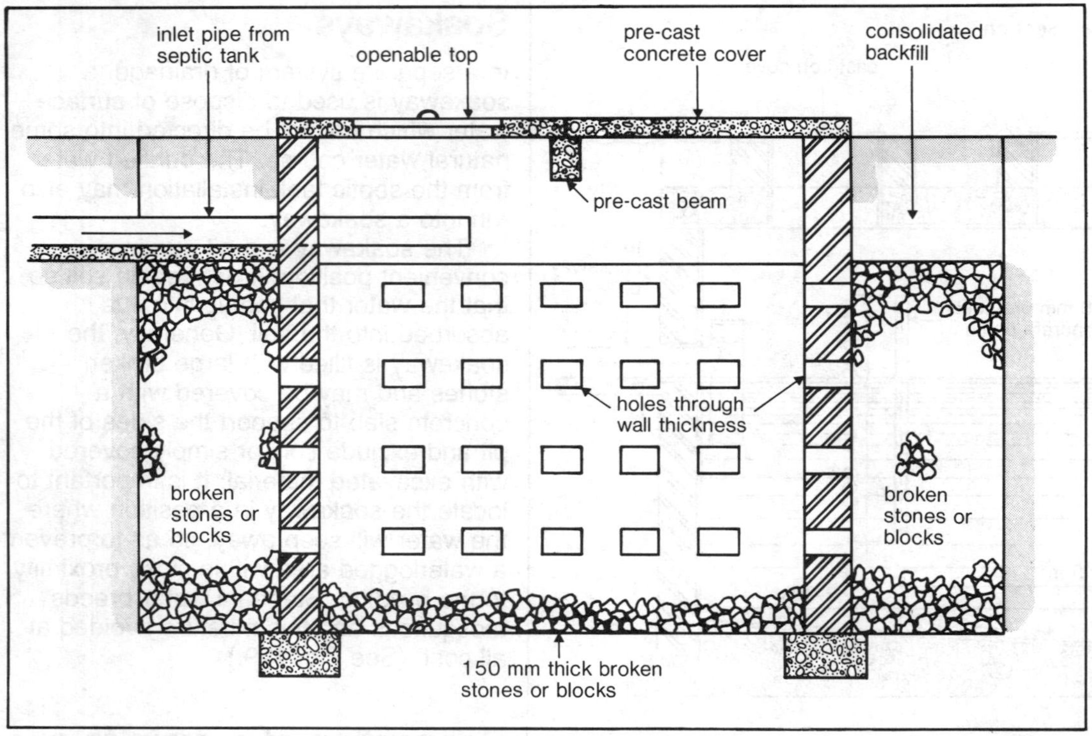

Fig. 199 Soakaway pit suitable for single family use

sand mortar. A stoneware saddle is illustrated in Fig. 200.

Testing of drains

Every drainage system must be tested to ensure watertightness, airtightness and quality of general finish of the system.

Drains can be tested in three ways as follows:

Hydraulic test

This is the most reliable method of testing drains. The test is performed by blocking the lower end of the section of drain to be tested by inserting an expanding plug or air inflated bag. The portion of the system is then filled with water up to the level of the gullies which should also be plugged or stopped with one of the devices made specially for this purpose. The filling with water is carried out by attaching one end of a length of rubber tubing to the nipple of the drain stopper or plug at the upper end and connecting the other end to a container holding water at the required head as shown in Fig. 201. A 600 mm head of water should be satisfactory. Entrapped air should be removed when applying this test.

A leakage is indicated by a drop in the level of water in the container, and as the drainpipes are haunched and covered the points of leakage are easily noticeable.

Fifteen minutes should elapse before the water level is observed, to allow for any absorption that might take place.

Smoke test

This test is performed by forcing air-laden smoke into the section of the system under test from a special smoke box, as shown in Fig. 202. The smoke is produced by burning oily waste in the smoke box. The smoke is forced through the lower end of the drain and vent pipes. Soil pipes as well as all traps should be left unsealed until the smoke emerges from them; this ensures that the drain under examination is full of air-laden smoke.

A few strokes of the bellows of the smoke machine will set up a slight pressure in the system, and the dome over the smoke box should rise and remain in that position if there are no leakages.

The advantage in this system is that any smoke escaping through a leakage is easily visible.

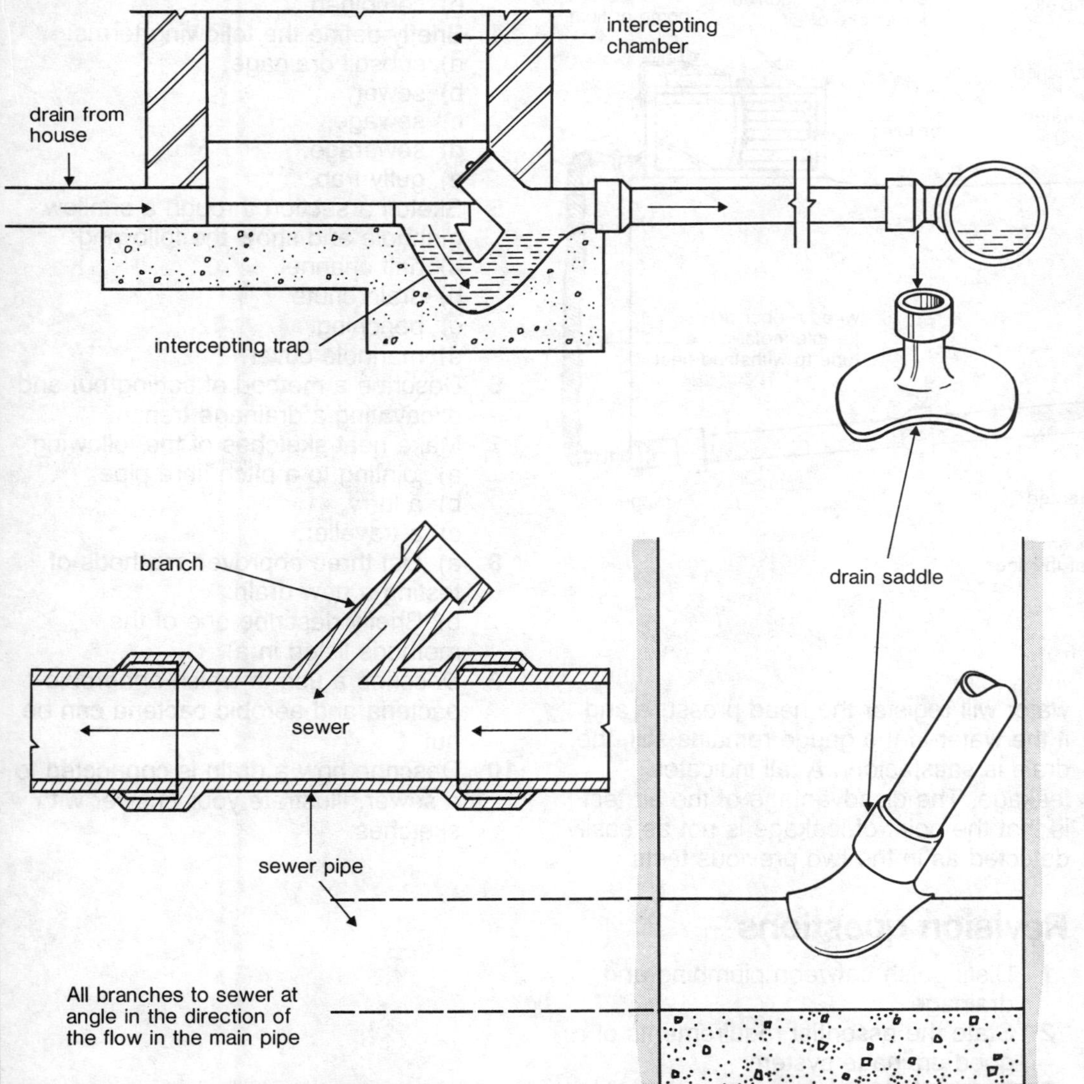

Fig. 200 Junction of drain and sewer

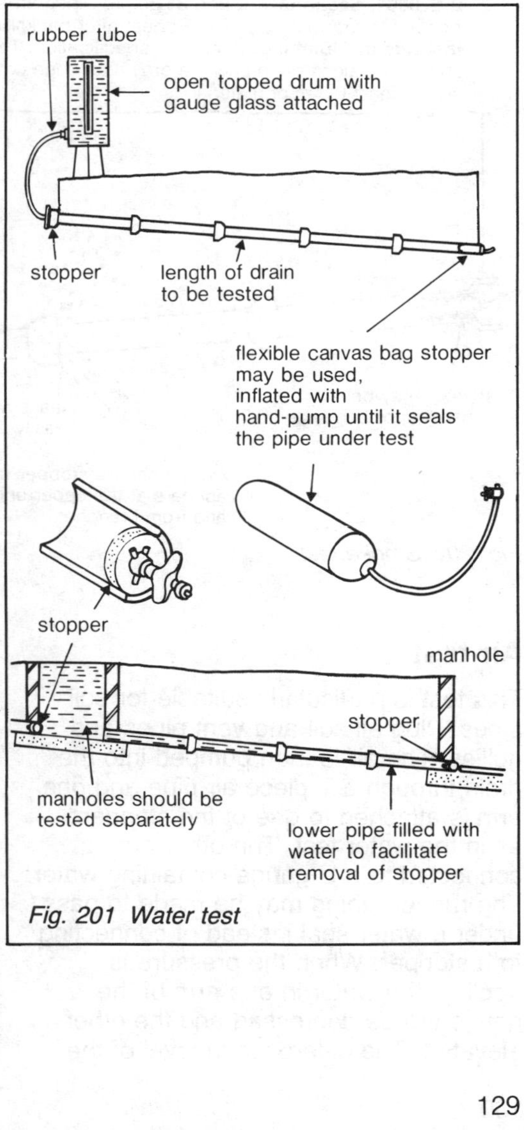

Fig. 201 Water test

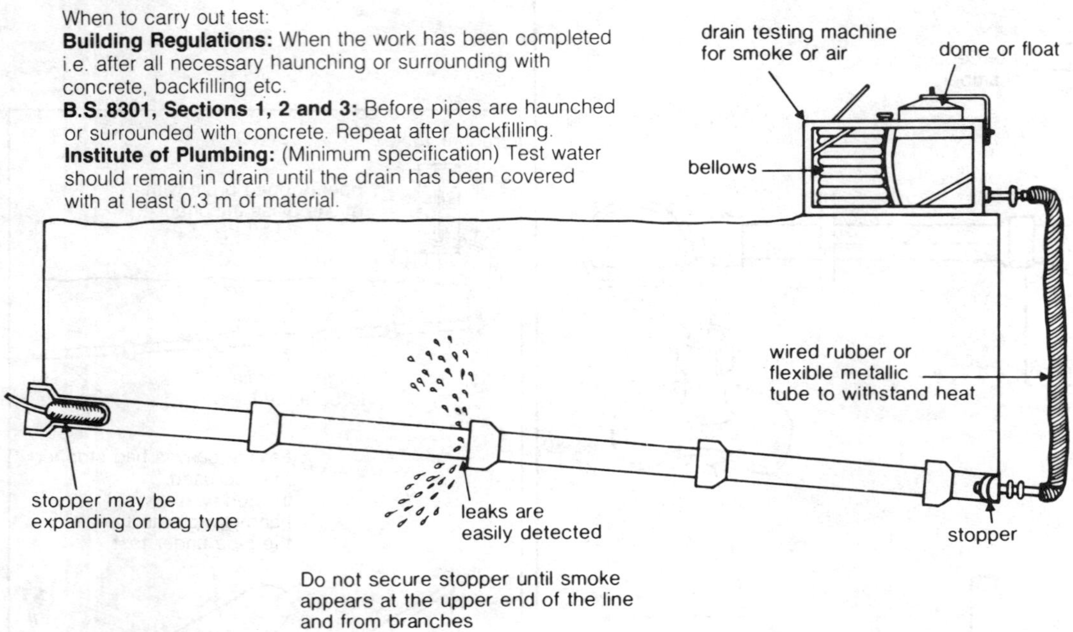

Fig. 202 Smoke test

Air test

This test is particularly suitable for soil pipes. Plug all soil and vent pipes and gullies. The air is then pumped into the drain through a T-piece air pipe and one arm is attached to one of the stoppers as in the water test. The other arm is connected to a U-gauge containing water. The rubber tubing may be made to pass under a water seal instead of connecting to a stopper. When the pressure is applied, the water in one arm of the gauge will be depressed and the other elevated. The difference in level of the water will register the head pressure and if the water in the gauge remains still, the drain is satisfactory. A fall indicates leakage. The disadvantage of the air test is that the point of leakage is not as easily detected as in the two previous tests.

Revision questions

1 Distinguish between plumbing and drainage.
2 State the essential requirements of a good drainage system.
3 Define the following systems of drainage:
a) separate,
b) combined.
4 Briefly define the following terms:
a) subsoil drainage,
b) sewer,
c) sewage,
d) sewerage,
e) gully trap.
5 Sketch a section through a shallow manhole and show the following:
a) half channel,
b) drain chute,
c) benching,
d) manhole cover.
6 Describe a method of setting out and excavating a drainage trench.
7 Make neat sketches of the following:
a) jointing to a pitch fibre pipe,
b) a larry,
c) a traveller.
8 a) List three approved methods of testing a new drain.
b) Briefly describe one of the methods listed in a).
9 Describe a use to which anaerobic bacteria and aerobic bacteria can be put.
10 Describe how a drain is connected to a sewer. Illustrate your answer with sketches.

16. Typical practical examination models

This chapter contains carefully designed examination models for workshop practice. These graded exercises are based strictly on the pattern of the City and Guilds of London Institute's practical tests in the trade. Models 1–6 are at the intermediate level while models 7–9 are at the advanced certificate level.

The building of the models should be conducted in typical examination conditions. Accompanying instructions should also be strictly followed. However, because of limitations in materials, workshop space and time in some institutions, trainees may be paired up while building the model and the time allowed halved.

Whenever space permits, a pictorial view of the model has been provided to aid full understanding. In order to obtain maximum benefit from these exercises, building them in the set conditions should be restricted to final craft trainees.

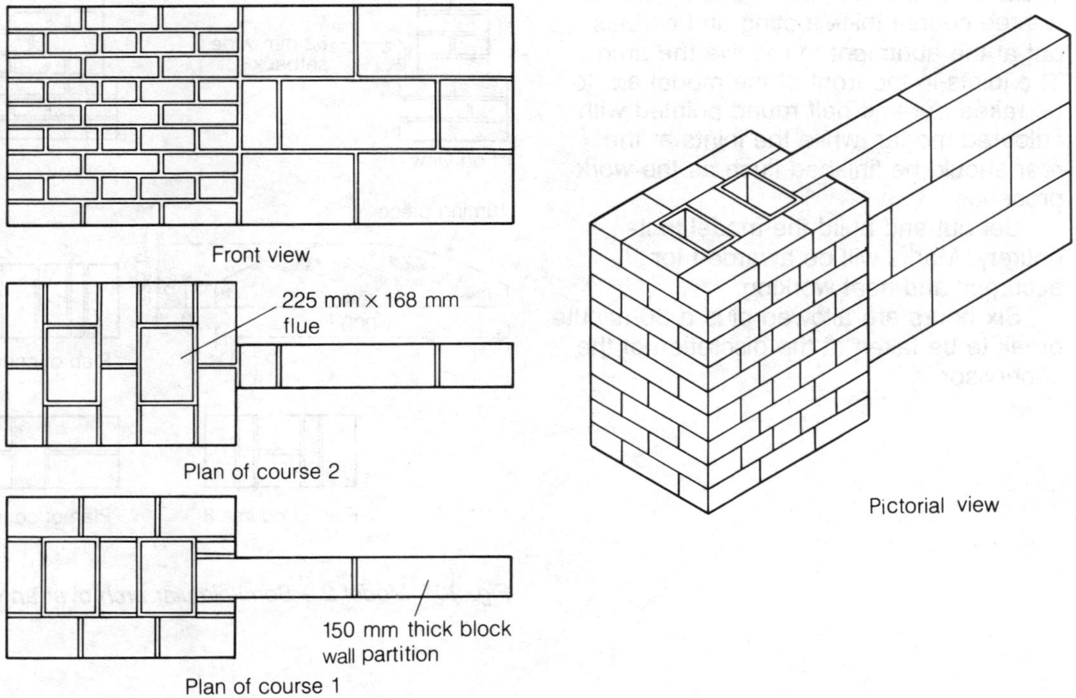

Fig. 203 Model 1 – Flue construction in brickwork

Intermediate level models

1 Flue construction in brickwork

The drawing shows a twin flue set in brickwork and a 150 mm thick solid block partition wall. The flues have internal measurements of 225 mm by 168 mm. They are to be pargetted as the work proceeds. The entire joints are to be raked out and weather struck pointed using a coloured mortar.

Set out and build the model in its entirety.

Six hours are allowed plus a 30-minute break to be taken at the discretion of the supervisor.

2 Semi-circular arch over an opening

The drawing shows an entrance to a housing estate spanned with a half-brick-deep segmental arch, carried on two bricks by one brick pillars. Each pillar has a three course thick footing and corbels out at the abutment to receive the arch. The joints in the front of the model are to be raked out and half round pointed with coloured mortar, while the joints at the rear should be finished flush as the work proceeds.

Set out and build the model in its entirety. Marks will be awarded for accuracy and neat working.

Six hours are allowed plus a 30-minute break to be taken at the discretion of the supervisor.

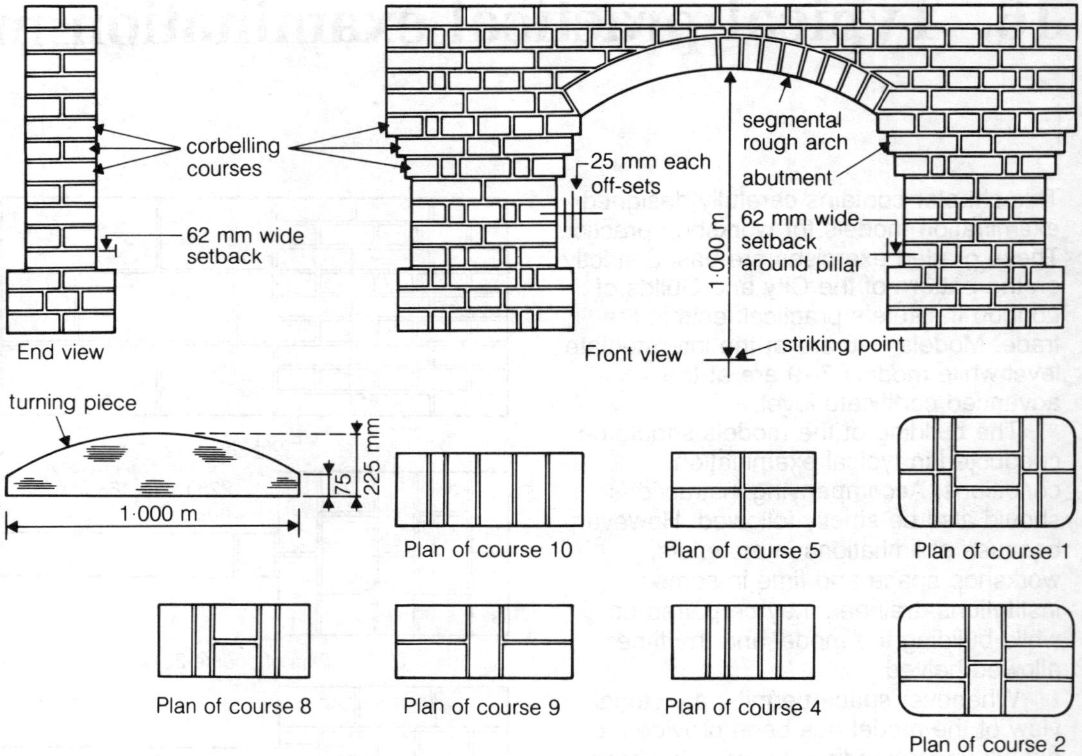

Fig. 204 Model 2 – Semi-circular arch over an opening

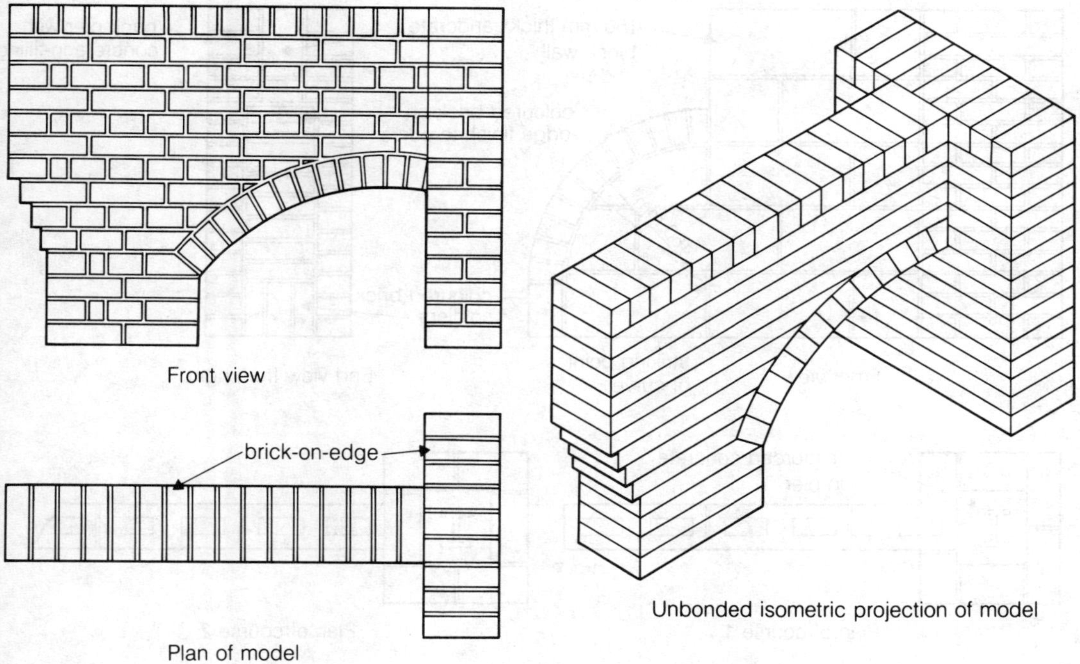

Fig. 205 Model 3 – Quadrant arch

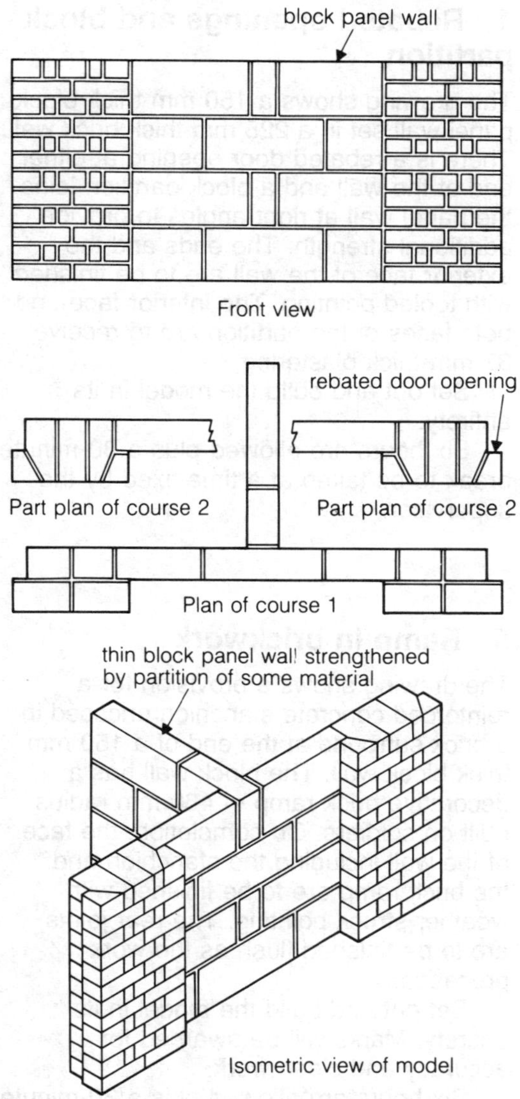

Fig. 206 Model 4 – Rebated openings and block partition

3 Quadrant arch

The drawing shows two one-brick party walls at right angles linked to each other by a quadrant rough arch over a passage in between the two walls. The model is bonded on the elevation in English bond. Produce plans one and two and fill in the correct bond. The faces of the two walls are to be pointed with tuck pointing on completion of the work while the rear joints are to be finished flush as the work proceeds.

Set out and build the model in its entirety. Marks will be awarded for accuracy and initiative.

Six hours are allowed plus a 30-minute break.

4 Rebated openings and block partition

The drawing shows a 150 mm thick block panel wall set in a 225 mm thick brick wall. There is a rebated door opening at either end of the wall and a block partition joins the panel wall at right angles to provide additional strength. The ends and the exterior face of the wall are to be finished with tooled pointing. The interior face and both faces of the partition are to receive 37 mm thick plastering.

Set out and build the model in its entirety.

Six hours are allowed plus a 30-minute break to be taken at a time fixed by the supervisor.

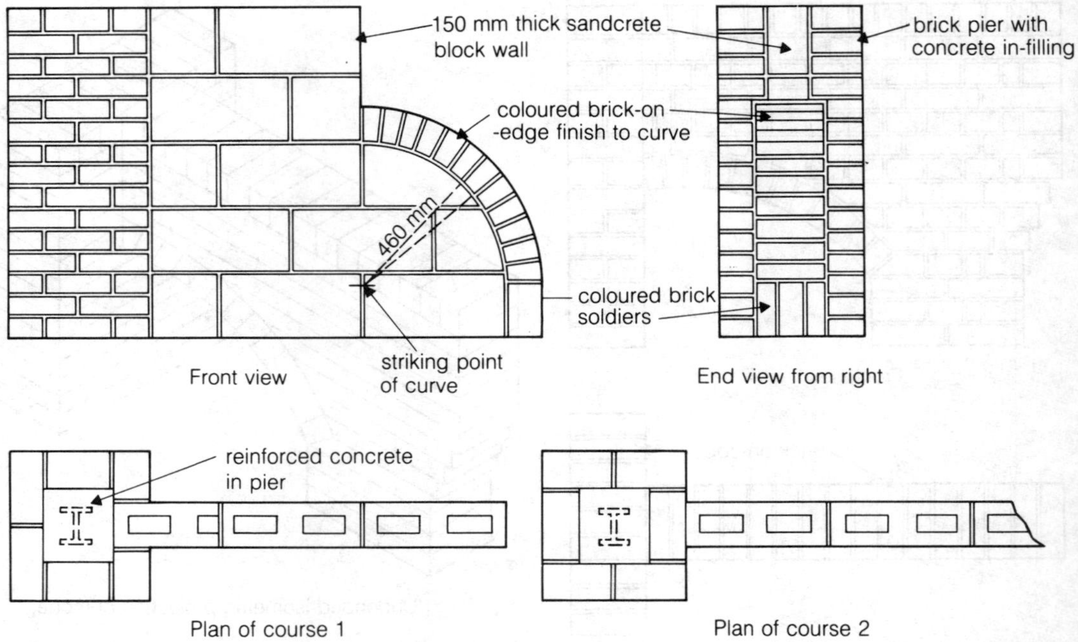

Fig. 207 Model 5 – Ramp in brickwork

5 Ramp in brickwork

The drawing shows a provision for a reinforced concrete stanchion encased in a brick surround at the end of a 150 mm thick block wall. The block wall has a decorative brick ramp of 460 mm radius built on soldiers. On completion, the face of the wall including the stanchion and the brick ramp are to be finished with weather struck pointing. The rear joints are to be finished flush as the work proceeds.

Set out and build the model in its entirety. Marks will be awarded for accuracy and clean work.

Six hours are allowed plus a 30-minute break to be taken at the discretion of the supervisor.

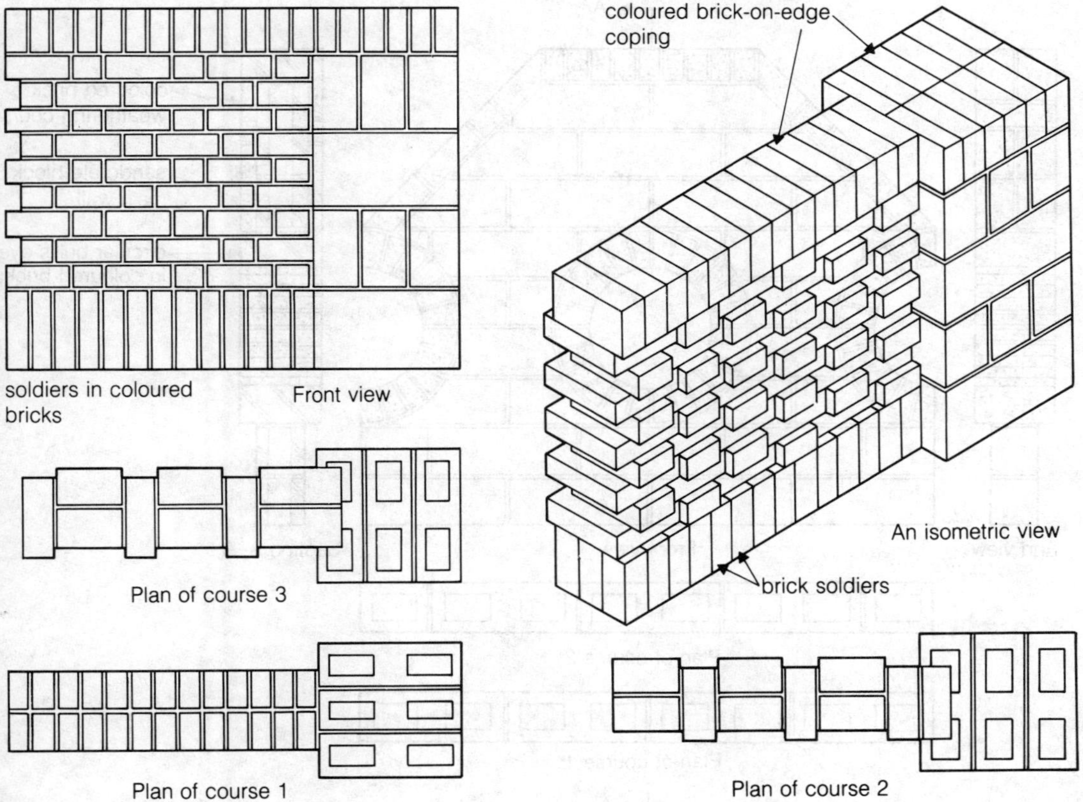

Fig. 208 Model 6 – Laced wall panel

6 Laced wall panel

The drawing shows a brick panel wall between block piers. The panel is built on brick soldiers base course. It is in Flemish bond with the headers projecting out and finished with a brick-on-edge course. The soldier course, headers in the wall and the brick-on-edge top course are built with decorative bricks. The base and top courses and the pier joints are to be finished with square recessed pointing. The rear of the model is to be patched up flat and plastered.

Set out and build the model in its entirety.

Six hours are allowed plus a 30-minute break to be taken at the discretion of the supervisor.

Advanced level models

The following models are typical examination questions for the advanced craft certificate. For this level of work, greater accuracy and a high class of craftsmanship are expected.

7 Bull's-eye or wheel arch construction

The drawing shows a part of a 150 mm thick block fence wall with bull's-eye or wheel arch built of decorative bricks 230 mm deep. The block wall is finished with brick-on-edge coping. The joints on the face and ends of the model are to be raked out and finished with half round pointing. The rear is to be plastered.

Set out and build the model in its entirety.

Six hours are allowed plus a 30-minute break to be taken at the discretion of the supervisor.

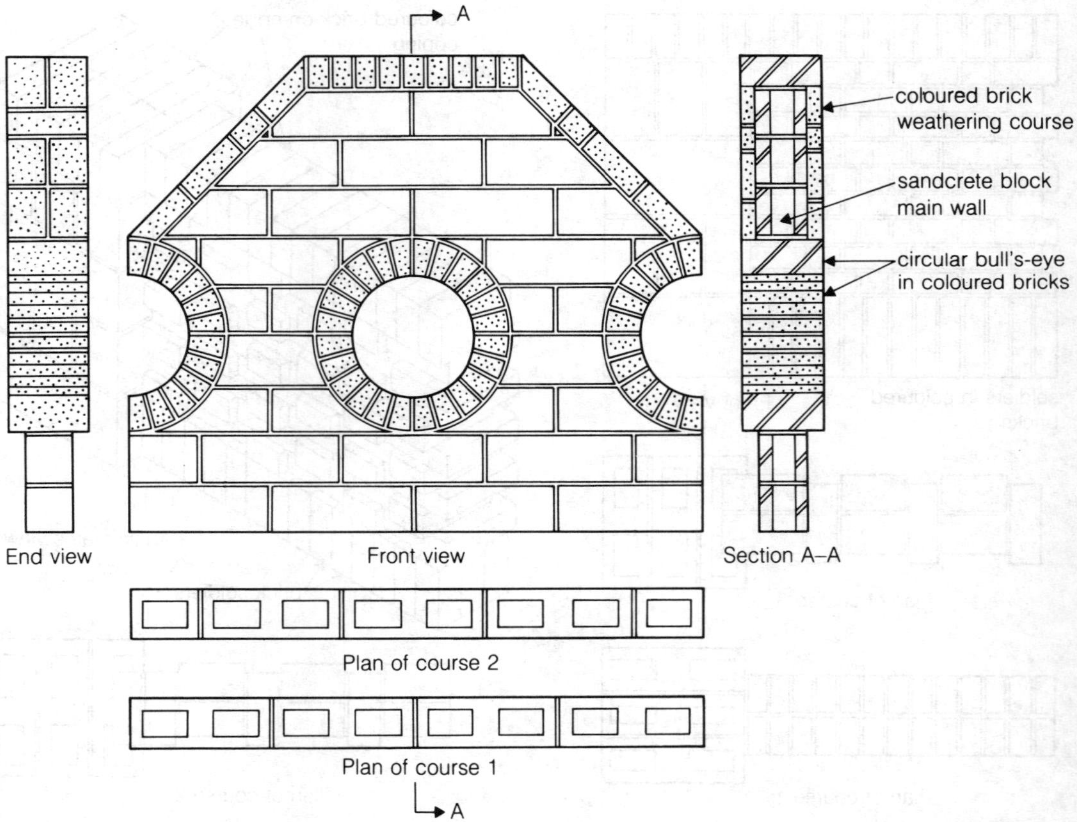

Fig. 209 Model 7 – Bull's eye or wheel arch construction

8 Stack bonded construction

The drawing shows a 150 mm thick block wall with two 1½ by ¾ brick attached piers of decorative bricks and a decorative brick stack bonded soldiers panel, built specifically to receive a plaque to be mounted on top. The piers have a string course projecting 56 mm from the general face and the panel is finished with a course of brick-on-edge as decorative features. The front of the model is to be finished with weather pointing while the rear is to be plastered.

Set out and build the model in its entirety. Marks will be awarded for accuracy, neatness and good craftsmanship.

Six hours are allowed plus a 30-minute break to be taken at the discretion of the supervisor.

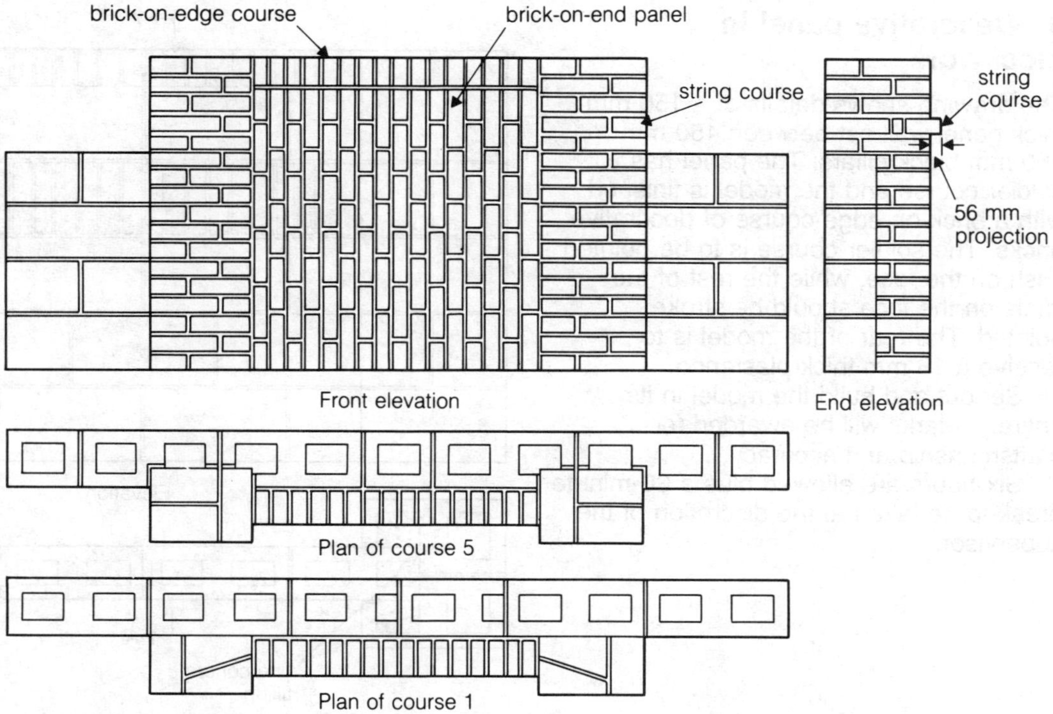

Fig. 210 Model 8 – Stack bonded construction

9 Decorative panel in blockwork

The drawing shows details of a 150 mm thick panel wall set between 450 mm × 450 mm block pillars. The panel has a soldier course and the model is finished with a brick-on-edge course of decorative bricks. The soldier course is to be pointed flush on the face, while the rest of the joints on the face should be stroke pointed. The rear of the model is to receive a 18 mm thick plastering.

Set out and build the model in its entirety. Marks will be awarded for craftsmanship and accuracy.

Six hours are allowed plus a 30-minute break to be taken at the discretion of the supervisor.

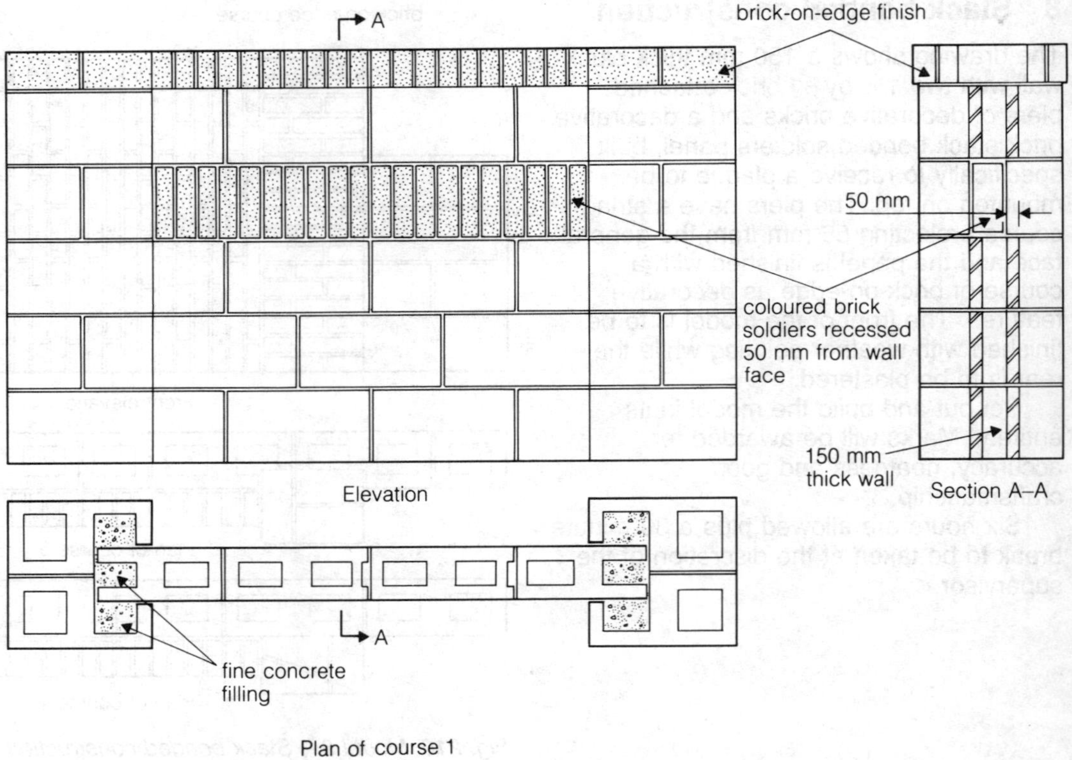

Fig. 211 Model 9 – Decorative panel in blockwork

17. Past examination questions

The following questions are from the City and Guilds of London Institute the West African Examinations Council, for both the intermediate and advanced craft certificate levels. These examination questions will provide useful preparatory aid for final classes entering for these and similar examinations. All seriousness should be given to their practice in the workshop, observing the instructions contained in the questions.

Intermediate level examination questions

1 Boundary wall

The drawing (Fig. 212) shows part of a boundary wall finished with a brick-on-edge coping. The three courses of block wall of the front elevation and the T-junction are to be built with the blocks (450 mm by 100 mm by 225 mm) provided. The remaining walls are to be built with the facing bricks. The top of the walls are to be built and finished with the facing bricks of contrasting colour to form the brick-on-edge coping. On completion, the joints in the front elevation are to be raked out and pointed with a half round jointing tool. The rear of the walls are to have a 12 mm

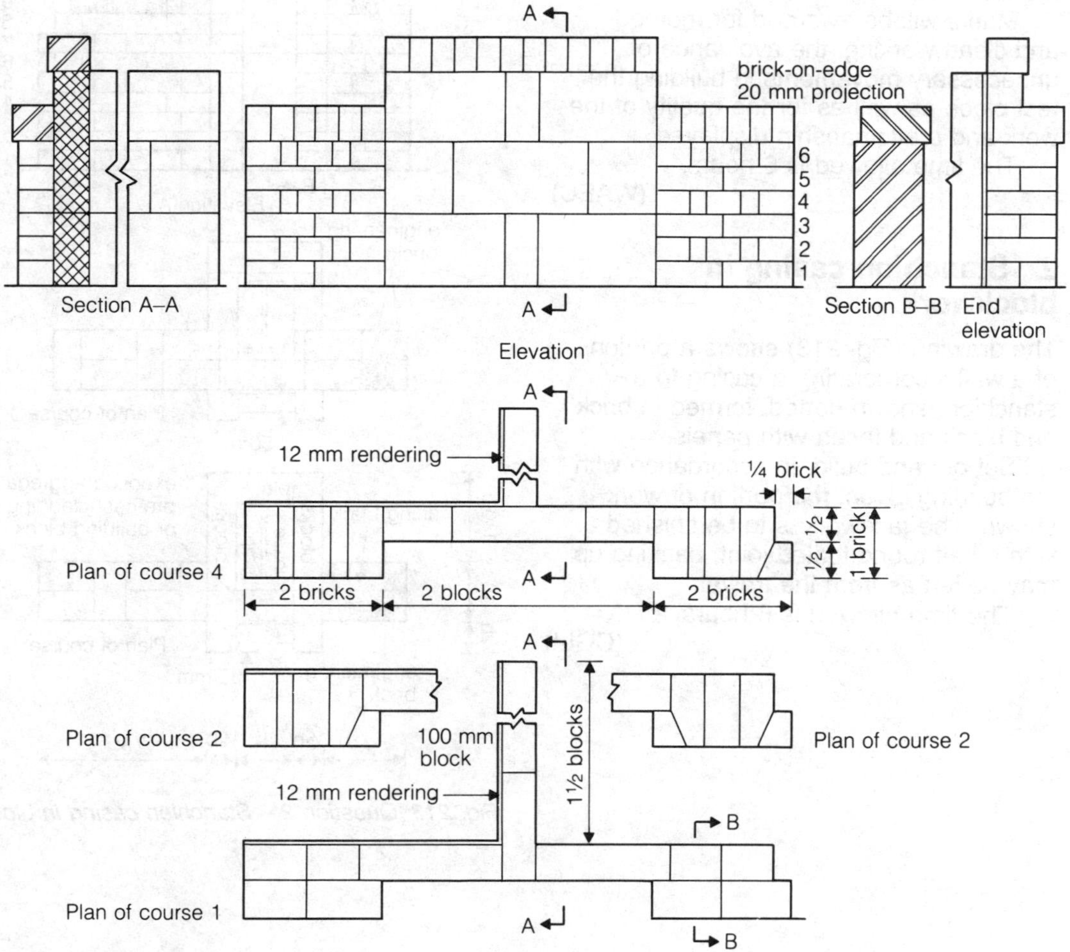

Fig. 212 Question 1 – Boundary wall

coat of rendering as indicated, finished with a steel floating trowel. The remaining joints to the back of the wall are to be left flush as the work proceeds.

The candidate is to set out and build the test piece in its entirety.

Marks will be awarded for method and clean working, the avoidance of unnecessary movements in building the test piece as well as for the quality of the work and craftsmanship displayed.

The time allowed is 6 hours.

(WAEC)

2 Stanchion casing in blockwork

The drawing (Fig. 213) shows a portion of a wall incorporating a casing to a stanchion, shown dotted, formed in brick and block and faced with panels.

Set out and build, in accordance with the bonding given, the portion of work shown. The facework is to be finished with a half round tooled joint, backing up may be left as from the trowel.

The time allowed is 6 hours.

(CGLI)

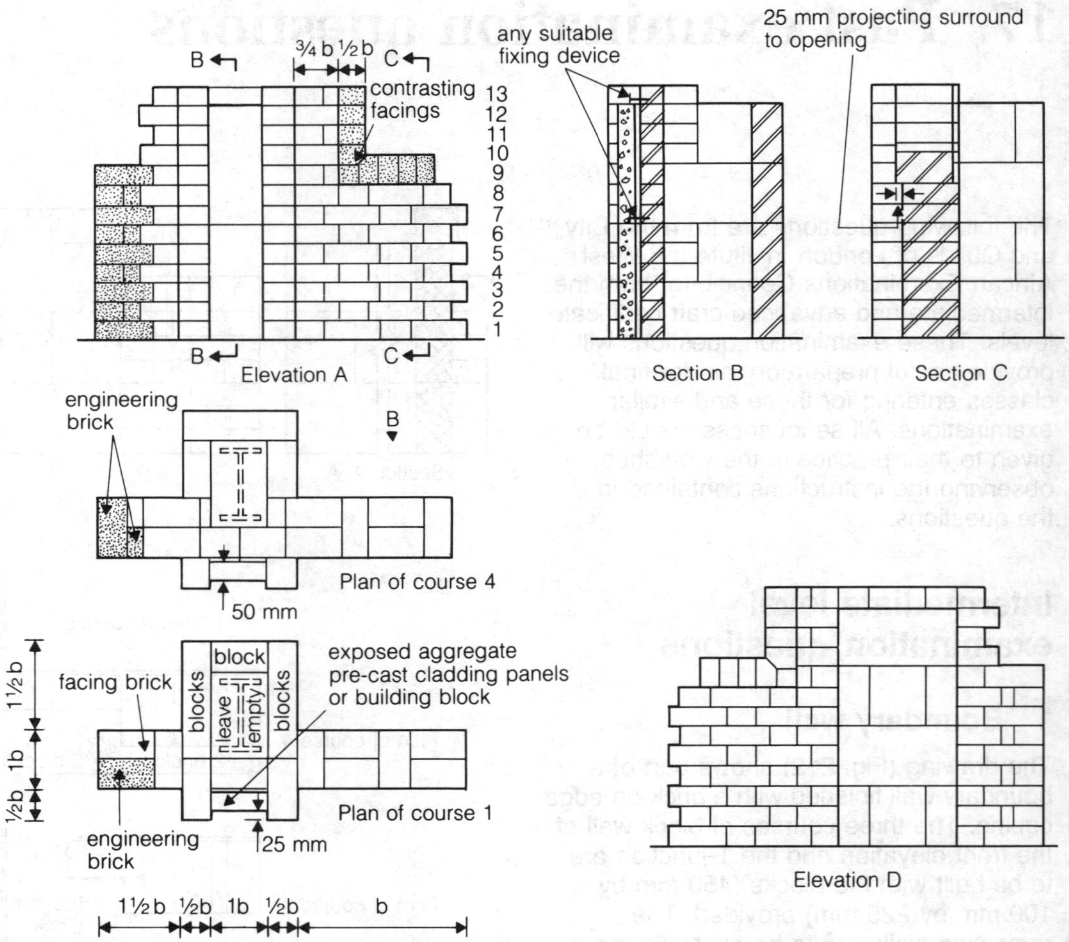

Fig. 213 Question 2 – Stanchion casing in blockwork

3 Basket weave panel

The drawing (Fig. 214) shows part of a boundary wall incorporating a basket weave panel and brick-on-edge paving course forming the base for a flower pot. The recess formed by the 112 mm thick block wall is to receive a basket weave panel and paving course as shown, using the contrasting bricks provided. The back elevation of the wall is to receive a 12 mm one coat rendering with a wood float finish. All joints to the face of the panel and paving are to be finished with a half round jointing tool on completion. Other joints on the face of the block wall are to be left flush as the work proceeds.

The candidate is to set out and build the test piece in its entirety.

Marks will be awarded for the method and clean working, the elimination of unnecessary movement in building the test piece as well as for the quality of the work and the craftsmanship displayed.

The time allowed is 6 hours.

(WAEC)

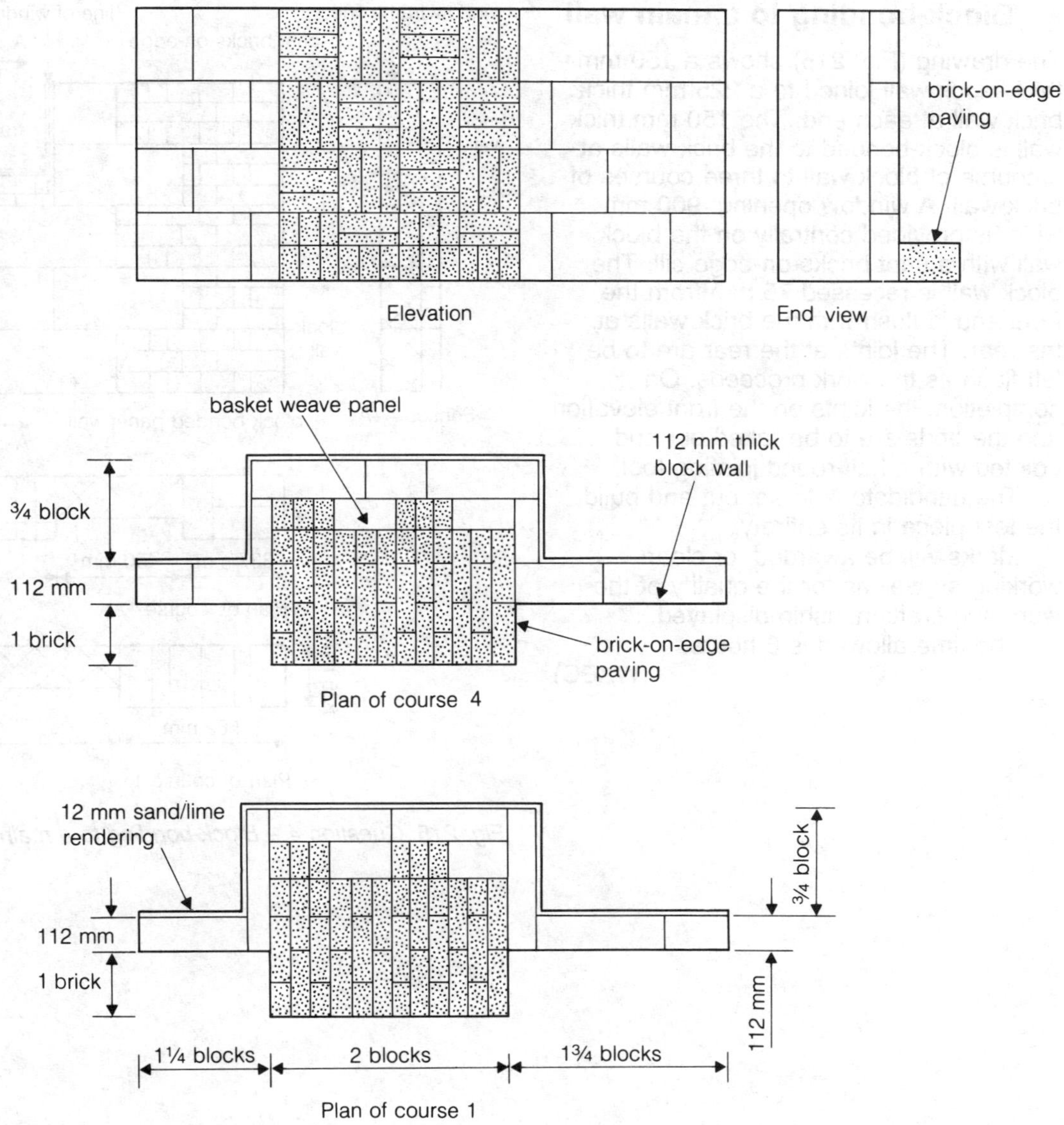

Fig. 214 Question 3 – Basket weave panel

4 Block-bonding to a main wall

The drawing (Fig. 215) shows a 150 mm thick block wall joined to a 225 mm thick brick wall at each end. The 150 mm thick wall is block-bonded to the brick walls at a course of block wall to three courses of brick wall. A window opening, 900 mm wide, is provided centrally on the block wall with a cant bricks-on-edge cill. The block wall is recessed 75 mm from the front and is flush with the brick walls at the rear. The joints at the rear are to be left flush as the work proceeds. On completion, the joints on the front elevation and the ends are to be raked out and pointed with a half round jointing tool.

The candidate is to set out and build the test piece in its entirety.

Marks will be awarded for clean working as well as for the quality of the work and craftsmanship displayed.

The time allowed is 6 hours.

(WAEC)

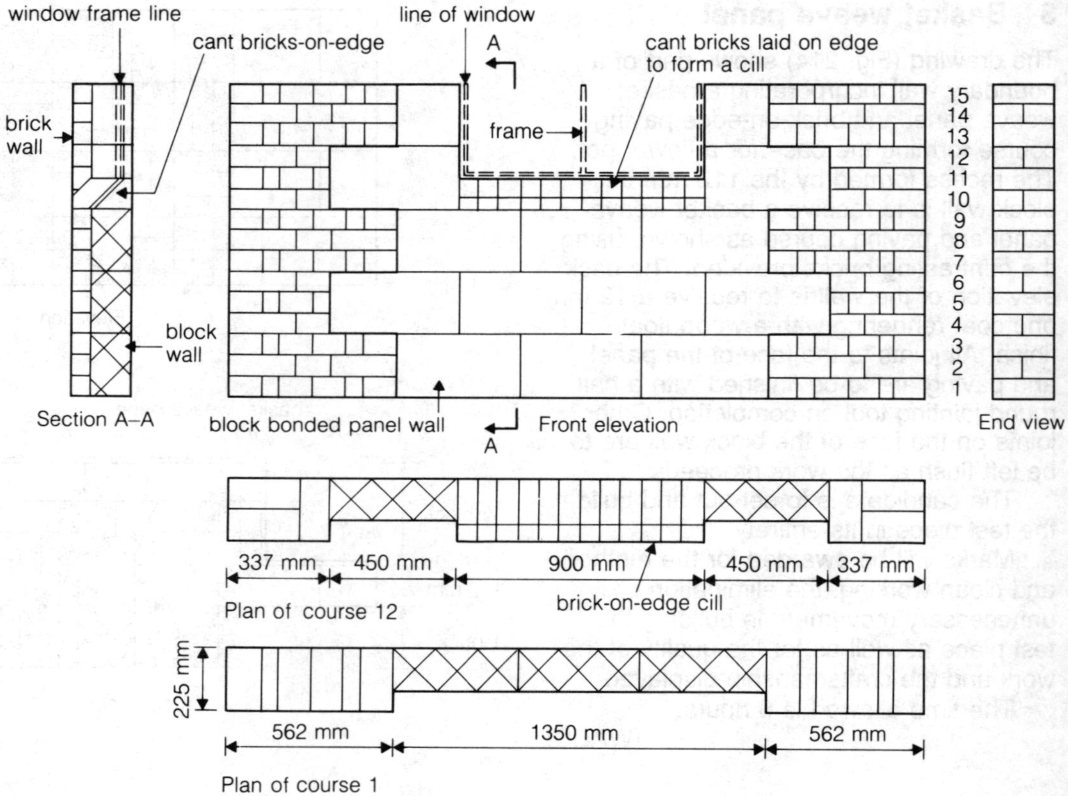

Fig. 215 Question 4 – Block-bonding to a main wall

5 Block piers and a panel wall

The drawing (Fig. 216) shows part of a boundary wall with attached block piers at both ends. The main brick wall is to be built in English bond and with common brick. The herringbone panel and the brick-on-edge copings are to be built with the coloured bricks. The attached block piers are to be built with the 450 mm by 225 mm by 150 mm blocks. On completion, the joints in the front elevation and the remaining attached block piers are to be raked out and pointed with a square jointing tool. The rear of the brick wall is to have a 10 mm thick coat of rendering to finish at the course below the brick-on-edge coping.

The candidate is to set out and build the test piece in its entirety.

Marks will be awarded for method and clean working, and the avoidance of unnecessary movement in building the test piece, as well as for the quality of work and craftsmanship displayed.

The time allowed is 6 hours.

(WAEC)

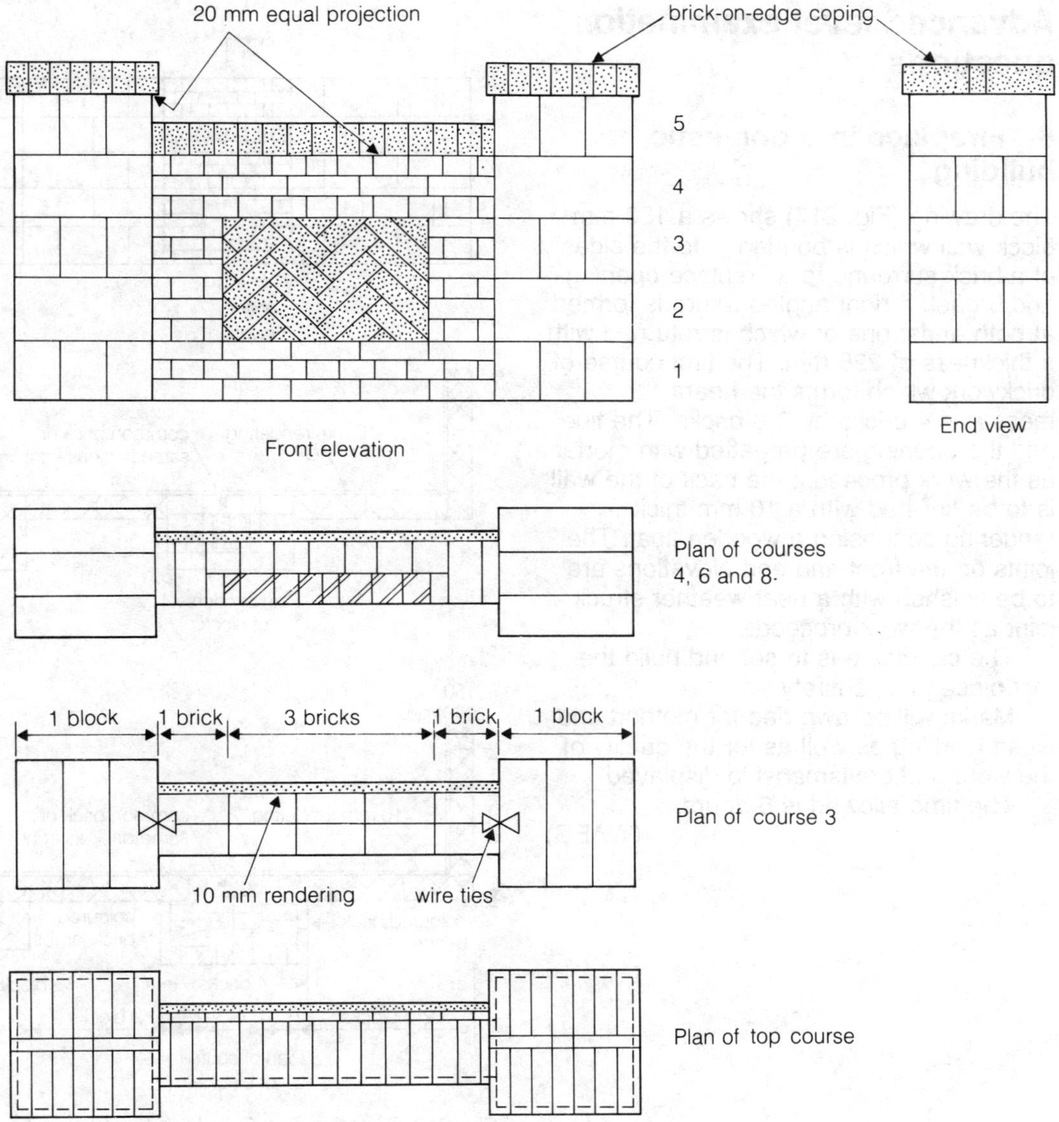

Fig. 216 Question 5 – Block piers and a panel wall

Advanced level examination questions

6 Fireplace in a domestic building

The drawing (Fig. 217) shows a 150 mm block wall which is bonded in to the sides of a brick surround to a fireplace opening and breast; a right angled return is formed at both ends, one of which is returned with a thickness of 225 mm. The first course of brickwork which forms the hearth measures 4 bricks by 2½ bricks. The flue and the opening are pargetted with mortar as the work proceeds; the back of the wall is to be finished with a 10 mm thick rendering coat using a wooden float. The joints on the front and end elevations are to be finished with a neat weather struck joint as the work proceeds.

The candidate is to set and build the test piece in its entirety.

Marks will be awarded for method and clean working as well as for the quality of the work and craftsmanship displayed.

The time allowed is 6 hours.

(WAEC)

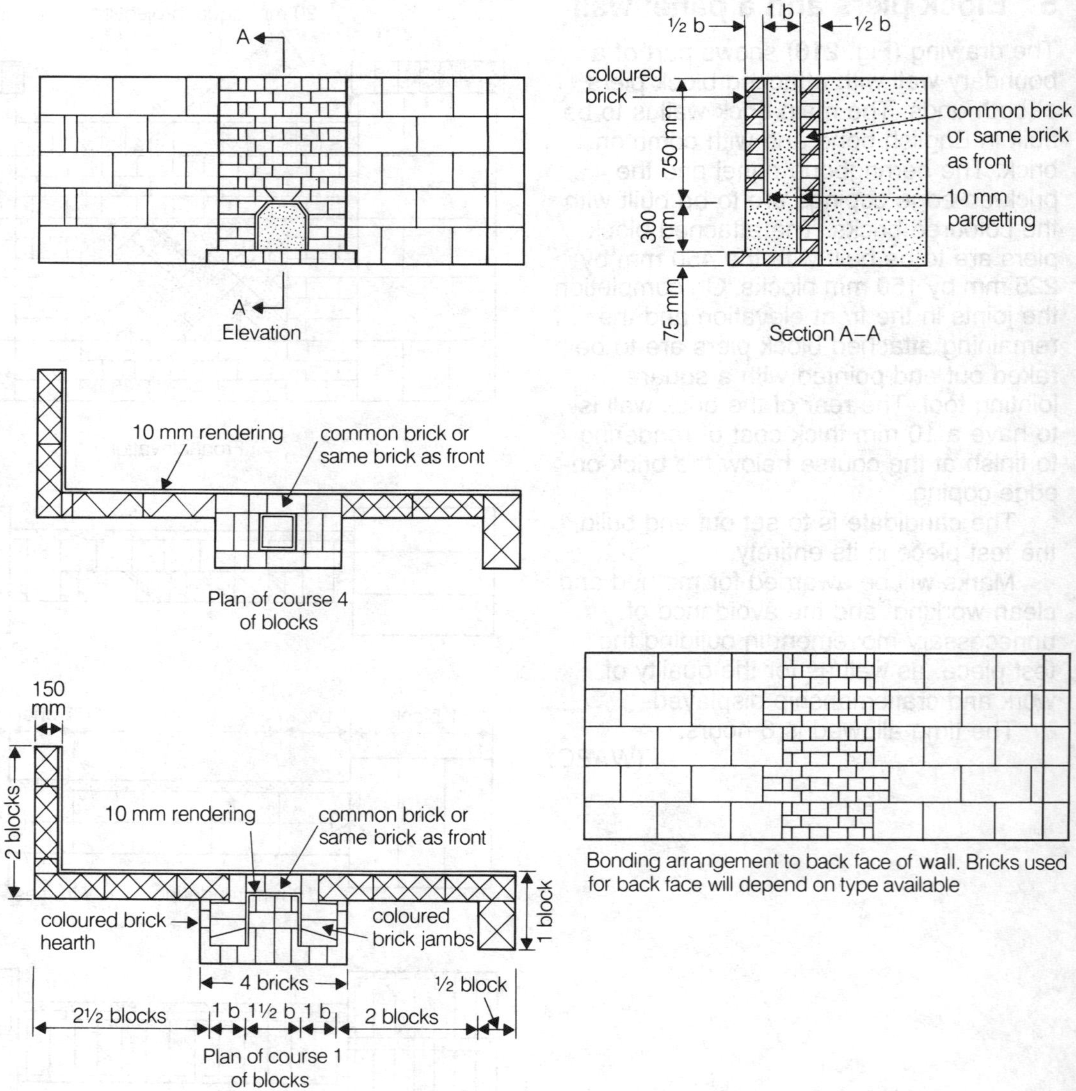

Fig. 217 Question 6 – Fireplace in a domestic building

7 Curved work in brick and blockwork

The drawing (Fig. 218) shows a two-brick square pier, being part of an entrance to a block of flats. A wing wall, 225 mm thick, and three and a half blocks in length, is attached to one side of the pier. This wall is to be finished with raking bricks. Attached to the rear of the pier is a curved wall, with a radius of 1·2 m built in header bond, which is to be rendered as shown. The remaining two faces of the pier have decorative features formed in coloured bricks. The top of the pier is to be finished with a protective capping and brick-on-edge coping on the wall. All joints on the faces and sides are to be finished with a neat weather struck joint on completion of the test piece. The back of the wall is to be left flush.

The candidate is to set out and build the test piece in its entirety.

Marks will be awarded for the method and clean working, and the avoidance of unnecessary movement in building the test piece, as well as for the quality of work and craftsmanship displayed.

The time allowed is 6 hours.

(WAEC)

8 Decorative panel wall

The drawing (Fig. 219) shows part of a sand/cement block boundary wall with decorative brick panel consisting of a soldier course at the base and overlaid with a basket weave pattern built to the

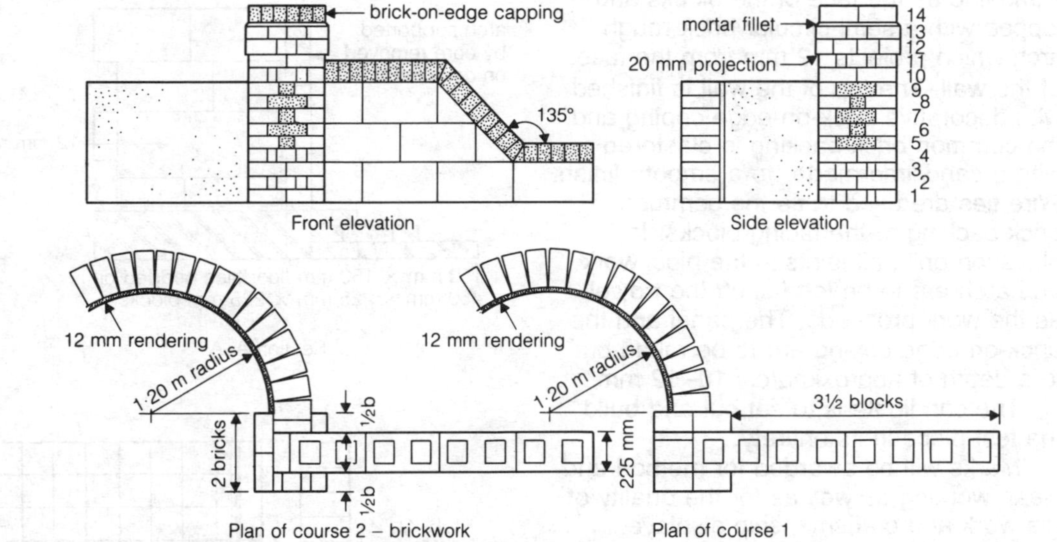

Fig. 218 Question 7 – Curved work in brick and blockwork

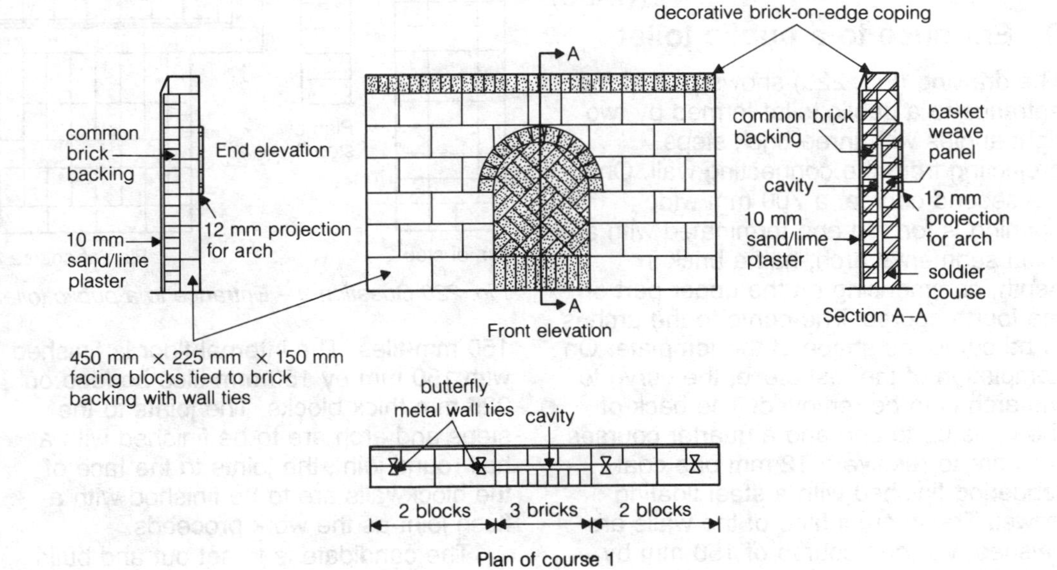

Fig. 219 Question 8 – Decorative panel wall

same line as the face of the blocks and topped with a semi-circular brick rough arch which projects 12 mm from the face of the wall. The top of the wall is finished with decorative brick-on-edge coping and the common brick backing is plastered with a sand/lime mortar to a smooth finish. Wire ties are used to tie the common brick backing to the facing blocks. In elevation only, all joints in the blockwork and arch are to be left full off the trowel as the work proceeds. The panel and the brick-on-edge coping are to be raked out to a depth of approximately 10–12 mm.

The candidate is to set out and build the test piece in its entirety.

Marks will be awarded for method and clean working as well as for the quality of the work and craftsmanship displayed.

The time allowed is 6 hours.

(WAEC)

9 Entrance to a public toilet

The drawing (Fig. 220) shows part of the entrance to a public toilet formed by two right angles with three brick steps projecting from the connecting wall. On the second course, a 700 mm wide opening is formed and terminated with an axed segmental arch, half a brick in depth, commencing on the upper part of the fourth course. The curve to the arch is to be cut to the shape of the template. On completion of the test piece, the curve to the arch is to be removed. The back of the walls up to one and a quarter courses high are to receive a 12 mm one coat rendering finished with a steel floating trowel. The internal face of the walls are finished with one course of 150 mm by

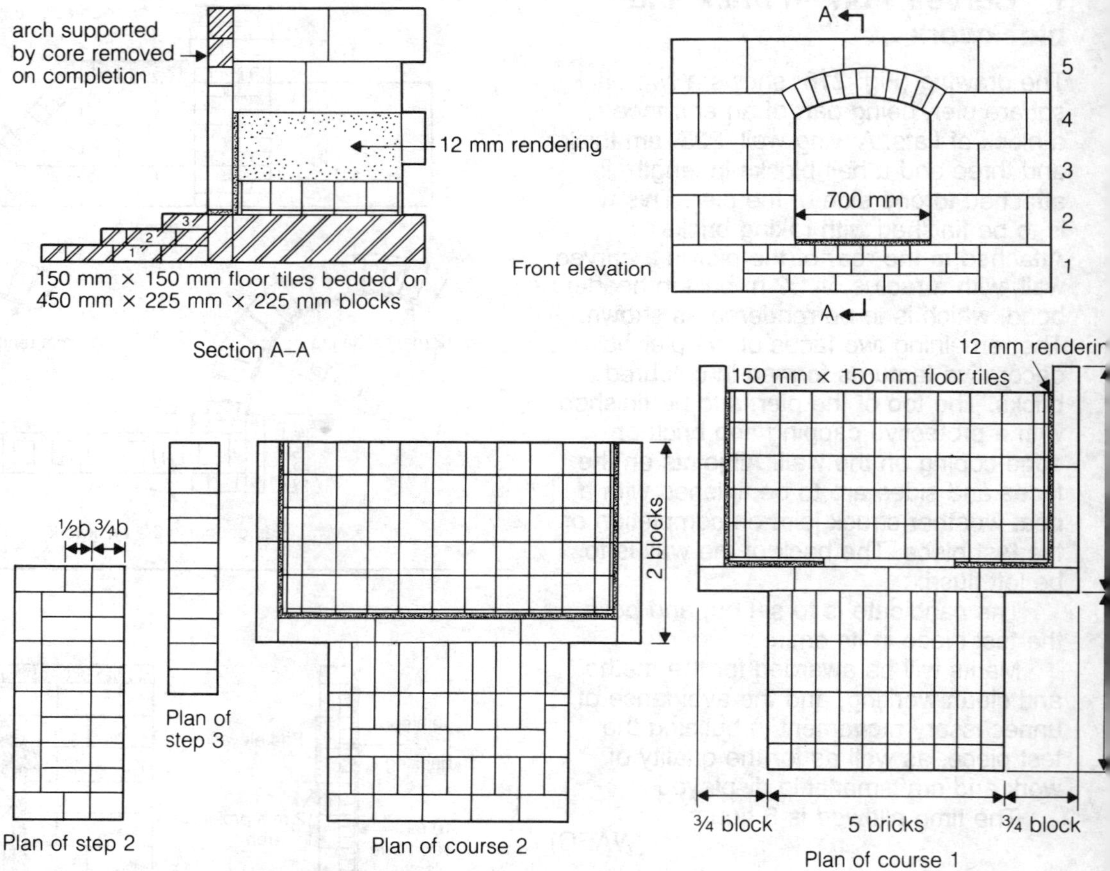

Fig. 220 Question 9 – Entrance to a public toilet

150 mm tiles. The internal floor is finished with 150 mm by 150 mm tiles bedded on 225 mm thick blocks. The joints to the steps and arch are to be finished with a half round joint, the joints to the face of the blockwalls are to be finished with a flush joint as the work proceeds.

The candidate is to set out and build the test piece in its entirety. Marks will be awarded for the method and clean working, the elimination of unnecessary movement in building the test piece, as well as for the quality of the work and craftsmanship displayed.

The time allowed is 6 hours.

(WAEC)

Glossary

Abutment End support designed to withstand the end thrusts of an arch or bridge.
Aggregates Loose particles such as broken stone, gravel or sand used together with cement to form concrete, mortar or screed, as the case may be. Aggregates are generally classified as fine and coarse.
Arris The sharp external angle of a block or brick formed by the meeting of two faces of the block.
Auger Cylindrical shaped metal with a cutting edge attached to a long handle used for boring holes in the ground.

Ballast Naturally occurring coarse aggregate used in the preparation of concrete.
Base plate A square flat steel plate with a central upstand piece used under the uprights of a scaffold.
Batch The quantity of concrete mixed at a time.
Batten A horizontal member fixed to main structural members on which plaster boards may be laid when constructing a lightweight timber partition.
Benching Method of sloping up the sides of half channels in a manhole or inspection chamber with cement mortar, to assist the flow of sewage and surface water.
Bitumen Mineral pitch or asphalt resembling tar, which does not crack or shrink. It is waterproof and durable.
Bituminous felt Felt impregnated with bitumen.
Boning rod A T-shaped wooden tool used for obtaining a rough level when laying kerb stones or preparing the bottom of a foundation trench to receive concrete.
Brace A diagonal member made of timber or steel fixed to prevent distortion of a framework.
Bridle piece A length of scaffold tube placed against the internal face of the wall to which other lengths, called ties, are put through the wall and tied, to strengthen the scaffold.
Building line A line established by the local authority, or agreed with the local authority, to which new buildings or structures must conform. The building line is established by the fronts of existing buildings along a street. The distance of the building line from the centre of the road varies according to the width of the road, i.e. whether is is a major or minor road.
Builder's square Large L-shaped and straight-edged woodwork with a diagonal frame piece used for setting out right-angled corners before the commencement of foundation work.
Buttress Attached pier of block or brick on the external wall of a building to provide additional support. Buttresses are sometimes built to batten into the wall.

Cast iron Metal manufactured by remelting pig iron together with scrap castings (if any) in a furnace. It is not normally malleable. It is an alloy of iron and carbon, the carbon being present as iron carbide or as graphite or a mixture of both.
Caulking A method of working tarred gaskin or hemp in between the socket and spigot of drain pipes with a special tool to centre the spigot in the socket.
Centering A curved wooden temporary support on which an arch is formed.
Chamfer Arris of a material cut off flat, usually at an angle of 45°. If the angle is not 45° it is known as a bevel.
Chase Groove formed or cut in a wall to accommodate cables, pipes or some other services.
Cladding A non-load-bearing external covering to the skeleton frame of a building or structure, such as a steel-framed building, with thin stone blocks, concrete slabs or sheets of metal.
Cleat A small piece of wood fixed to a structural member to help support another member, usually in making formworks.
Clinker In the manufacture of cement, the resultant from burning cement slurry. Generally, residue from an iron furnace; well-burnt bricks used for paving.
Commons Second rate bricks selected after burning, and used for non-facing works.
Composite wall A wall whose total thickness is composed of two different types of material.
Coping Concrete slabs, stones, bricks or slates laid on top of a wall as a finish.
Corbel Projection of a part of a building structure having a bearing in a wall to support a superimposed load.
Coupler Tubular scaffold fitting used to tie the various lengths together to form the scaffold. There are various types of coupler, namely, double coupler, single coupler and swivel coupler.
Course A row of blocks or bricks in a wall.
Cowley level Automatic levelling instrument.
Cramp a) Thin piece of metal bent at the ends used to hold two blocks permanently together.
b) A tool used to force boards or joists close together.
Crazing Fine hair-like cracks on plastering, rendering or floor finish.
Curing Method of keeping a newly laid concrete moist until it gains sufficient strength.
Cutting gauge A flat piece of wood having small square pieces nailed to it at distances equal to half bat and three quarter bat. Used to mark a brick before cutting.

Damp proof course (D.P.C) Any material which dones not permit the movement of moisture through its mass, and is used to prevent the entry of dampness into a building.
Damp proof membrane (D.P.M) A thin sheeting of damp proofing material, e.g. bituminous felt.
Datum peg A peg on top of which is marked the datum level from the nearest permanent Ordnance Bench Mark plus or minus level figures from the architect's or engineer's drawings. It is used as a point of reference and a guide to tradesmen on a building site.
Deadman Unjointed pier of blockwork erected to hold a levelling line. It is employed by the blocklayer to maintain the plumb and flatness of a gable wall. The deadman is built to the same gauge as that of the permanent wall.
Double bullnose brick A brick having the top arrises at both ends rounded.
Double cant Brick bevelled or slanted at both ends through the thickness.
Dowel Wooden pin for securing the ends or sides of two pieces of wood.

Drain saddle A drain joint with a specially shaped and enlarged end used to connect a branch drain to the sewer.

Extrados The external curved line of an arch.

Fireload Total quantity of flammable materials within a room compared to the volume of space in the room.
Flaunching Cement mortar worked round the tops of chimney stacks to form a weathering.
Floating Smoothing the surface of newly laid concrete with a wooden or steel trowel called a float.
Fork head Adjustable head attachment having two forks, which is mounted on a prop to receive a wooden beam or joist.
Foundation The lowest structural part of a wall or pier.
Frog A triangular or rectangular shaped depression on the top of a brick to reduce the weight.

Gable That part of a wall which follows the pitch of the roof from eaves to ridge and seals the end of the roof.
Gaskin Yarn rope used for the purpose of caulking joints in a drain-laying operation.
Gin wheel Wheel with an attachment whereby a scaffold tube may be placed (see section on scaffolding).
Grout A watery mix of cement poured into a joint to form a solid fill.

Half bat Half a brick cut across the width.
Hardcore Collection of loose inert particles such as broken blocks, bricks or stone placed to form a firm reliable base for concrete ground floors or similar constructions.
Haunching Method of protecting a newly laid and tested drain whereby concrete is placed against the side of the pipes and worked to slopes on either side.
Header The face of a block measuring 225 mm^2 or 225 mm by 150 mm.

Intrados The internal curved line of an arch.
Invert The lowest point in a pipe bore, or sewer or the lowest level of a drain trench.

Jamb Vertical side of an opening.
Jointing Finish given to the joints of a wall as the laying proceeds, using the same mortar as for laying, to give a neat appearance and prevent the penetration of moisture through the joints.
Joist Timber or steel place and supported at each end to carry a floor.

King closer Brick bevelled from the centre line of the header to half way along the stretcher face.

Laggings Narrow strips of wood fixed round the top edges of a wood centre for an arch to carry the voussoirs.
Larry (scraper) a) A flat semi-circular head of steel or wood attached to a long handle used to remove surplus jointing mortar from the internal bore of a drain pipe.
b) Plasterer's implement used for mixing coarse stuff.
Lath A piece of wood with straight edges.
Ledger The horizontal member which connects standards in a row.
Lintel The horizontal bridge over a window or door opening usually formed of reinforced concrete, block with concrete, or brick and steel. Treated timbers of suitable dimensions are used as lintels in rural buildings.

Manhole A chamber of blockwork or brickwork built on a drain line or sewer to give access for cleaning, inspection and repair. There now exist pre-cast concrete and glazed fireclay manholes which are quicker to install than building a traditional manhole.
Mass concrete Concrete of considerable thickness without reinforcements.
Mild steel Steel containing a small amount of carbon, about 0·25 per cent, which makes it softer than cast iron. Used as reinforcement in concrete.
Mitred bat A bevelled half a brick, measuring 56 mm by 75 mm on one side and 112 mm by 75 mm on the opposite side.
Mortar A plastic mixture of cement, sand, with or without lime, and water, which is used to bed building materials. Mortar, when set hard, binds building units together firmly.

Parapet A guard wall erected to hide the slope or pitch of a roof, or on a balcony or bridge.
Partition A wall whose primary function is to divide space within a building into rooms.
Party wall A common wall which separates two buildings or two flats in the same building.
Perpend Vertical joints in a wall.
Pier a) Supporting pillar for a beam, bridge or an arch, in steel, brick, block, stone or concrete.
b) Vertically protruding unit of similar materials formed on the face of a wall to give necessary additional support for point loads.
c) Any narrow vertical unit of brick, block or stone between openings.
Pinch rod Piece of batten cut to the width of an opening which is used to strut the foot of a door frame apart to ensure that the width of the door remains the same throughout.
Plaster To apply coats of cement and sand mortar to internal faces of walls to obtain a truly flat finish.
Plinth header Brick bevelled on the top; the bevel starts from about 25 mm from the base to about one-half or one-quarter of the length to the top. A plinth is used to reduce wall thickness and weather the wall.
Plinth stretcher A bevel along the stretcher face of a brick, starting at about 25 mm from the bed and terminating at the centre line of the brick.
Plumb bob A ball of lead suspended on a string. It is used for checking the plumb of blocklayer's work, such as rendering.
Pointing Finish given to the joints of blockwork or brickwork after the wall has been built to prevent the penetration of moisture through the mortar joints and to form a neat appearance.
Poling board The vertical member placed against the side of the trench in a timbering operation.
Profile Wooden rail used as a guide in setting out a trench for foundation work.
Purlin Horizontal supporting member on rafters on which the roof coverings are fixed.
Putlog A short length of scaffold tube with one flat end. Putlogs support the platform of a scaffold, having the flat ends pushed into the wall joints and the plain ends tied to the ledger.

Quarter bond A bond in which the bricks lap one-quarter of the total length over the last joints in the course directly below them. Bonds such as English and Flemish are known as quarter bonds.
Quarter lap Distance to which end of a brick laps over the last vertical joint in the course directly below it, as seen from the elevations of a wall built in English or Flemish bond.
Queen closer One half of a brick split along the stretcher face.
Quoin The external angle formed by meeting of two walls. The external corner of a wall.

Racking back Provision for future extension in which each course of the wall is set back.
Rafter An inclined main beam that carries the roofing.
Raft foundation A shallow foundation cast over the whole area of a building, used in sandy soils. It carries the load that is placed on it using the principle of a raft floating on water.
Ranging line Lines drawn from profile to profile showing the outlines of walls and concrete foundations when setting out a building.
Reinforced concrete Concrete in which steel rods

are laid during the casting to yield a high strength.
Render To apply coats of cement and sand mortar to external faces of walls to obtain a truly flat finish.
Reveal The square end of a door or window.
Reveal pin Piece of tubular stee with one end having a screwed steel adjustable head, both ends being finished with flanges. When placed horizontally in a window or door opening, it can be tightened against the reveals by screwing the adjustable heads.
Rise The perpendicular line between the springing line and the lower end of an arch crown.
Rodding arm Branch in an intercepting trap purposely formed as an access for cleaning the drain pipe.
Rubble Any stonework which is not highly finished. There are various types, from random rubble to regular rubble.
Rustic Roughened or uneven surface.

Sand blinding A method of spreading the surface of hardcore with coarse sand or other similar material and rolling it in to fill up spaces in between the hardcore, prior to concreting.
Screed a) Semi-dry mixture of sand and cement used either as a base to receive a floor finish or as a topping for a concrete floor.
b) A strip moved over a guide to strike off or finish a surface.
c) A straight-edge used for striking off or finishing a surface.
Segregation The separation of the heavier materials in a wet concrete mix during transportation and pouring.
Septic tank A chamber of sewage treatment suitable for areas without public or private sewers.
Setting out Process of marking out the outlines of a proposed building on the ground, also, marking out the outlines of a wall on a concrete foundation.
Sewer A large drain which receives and conveys away waste from houses and streets to a place of disposal.
Shoring jack A straight steel tube attached to a flat-ended screwed adjustable head used as a prop in decking formwork.
Sill (cill) The horizontal finish given to the foot of a window opening. External window sills are usually projected out and throatings provided under the projection to drip rainwater off the wall. A sill may be of concrete, brick, tiles or wood.
Single bullnose brick A brick having the top arris rounded to a quadrant at one end.
Single cant Brick bevelled or slanted at one end through the thickness.
Sleeper wall Short wall built within a hollow floor of timber to support the floor joists. The wall is often built with holes through it for free circulation of fresh air within the floor, in which case it is called 'honey-combed sleeper wall'.
Slump cone Bucket-like metal container used in carrying out a slump test to measure the workability of freshly mixed concrete.
Slurry In the manufacture of cement, the paste-like mixture of limestone and clay before it is fired in a kiln.
Snecking A regularly curving feature of blockwork in either direction.
Soakaway pit A pit dug close to a septic tank and filled with broken blocks or stones into which sewage from the septic tank discharges and later soaks away into the soil.
Soffit Under side of an arch, lintel, cornice or eaves.
Span Distance between the reveals of an opening.
Spatter-dash Mixture of coarse aggregate or rough sand and cement, applied to wall surface as a finish on its own, or to provide a good key for rendering coats.
Spigot The threaded end of a drain-pipe for fixing into a socketed end of the next pipe in a drain-laying operation.
Spirit level A wooden or metal straight-edge with inset glass tubes containing spirit and an air bubble. When the straight-edge is placed on an object which is level, or plumb, the air bubble moves to the centre of the glass tube.
Springing level A level from which an arch begins to curve. It is also referred to as the springing line.
Squint Oblique angled brick or opening formed in a wall. A squint angle may be acute or obtuse.
Standard An upright member of a scaffold positioned in alignment and parallel to the face of the wall.
Stirrup Shear reinforcement which passes round and secures to the main reinforcement.
Stretcher The longer face of a block measuring 450 mm by 225 mm.
Striking point Point from which the radius line of an arch starts.
Strut A short horizontal member placed across a trench to hold the timbering at both sides of the trench firmly apart. It carries compressive forces.
Suction a) Partial vacuum by wind which tends to lift a roof covering.
b) Absorption of moisture from blocks, plaster or paint due to the porous surface to which it is applied.

Tamping rod Piece of steel rod of standard dimension used in tamping the concrete in test cube preparation, or in a slump test.
Tanking A method of waterproofing a basement in which the vertical damp proof course is joined to the horizontal one.
Terrain A stretch of undeveloped land.
Three-quarter bat Three-quarters of a brick cut across its width.
Throating Narrow groove formed under the front edge of stone or concrete coping, concrete balcony or sill to prevent water penetration into the wall.
Tingle plate Flat sheet of metal with the end cut into three fingers through which a line is passed to hold it tight and prevent sagging at the middle. It is used in building a long wall.
Toe board An inclined plank secured to the standards of a scaffold to prevent loose materials falling off the platform and, as an additional safety guard, for the feet of persons using the scaffold.
Tongued and grooved boarding A system of joining boards of timber floor grooved on one side and tongued on another side. To join the boards, therefore, the grooved side is pressed on to the tongued side of another timber and cramped home.
Toothing The recesses formed in alternate courses of a wall to facilitate future extension to the wall.
Topography Description of the features of a stretch of land which shows rivers, valleys, high ground, roads, etc.
Transom A short piece of steel tube with both ends plain, used in the same way as a putlog.
Transom clip A 90°C fitting which ties a transom to a reveal pin in an opening.
Traveller T-shaped wooden tool used in laying drainpipes and also for obtaining the correct slope during the excavation of the trench.
Trench A narrow cutting in the ground into which the foundations for structural walls are laid.

Vent Duct or pipe discharging unwanted air from any system.
Voids Air spaces within a mass of concrete, hardcore or aggregate.
Voussoir The building units comprising the arch, which may be of bricks, specially cast concrete units or purpose-made blocks.

Waling board A horizontal member placed against the side of a trench to hold poling boards in position.
Wall plate Horizontal timber bedded on top of the wall to which the floor joists or rafters are fixed.
Wall-tie Specially shaped metal wire with a central drip, built across the air duct in a cavity wall, so as to drip any moisture that may accummulate on it.

Y-junction Drain joint connecting two branches meeting at a point.

Index

Note: References in *italic* are to illustrations. The same numbers may also refer to text.